A Brief History of the Future

The origins of the internet

John Naughton

Weidenfeld & Nicolson
LONDON

First published in Great Britain in 1999
by Weidenfeld & Nicolson

© 1999 John Naughton

A CIP catalogue record for this book
is available from the British Library.

ISBN 0 297 64330 4

Typeset at The Spartan Press Ltd,
Lymington, Hants
Printed in Great Britain by
Clays Ltd, St Ives plc

Weidenfeld & Nicolson
The Orion Publishing Group Ltd
Orion House
5 Upper Saint Martin's Lane
London, WC2H 9EA

To the memory of Vannevar Bush,
who dreamed it up, and to Tim
Berners-Lee, who made it happen.

Contents

Preface

For all knowledge and wonder (which is the seed of knowledge) is an impression of pleasure in itself.

Francis Bacon, *The Advancement of Learning*, 1605

History, someone once said, is just the record of what one age finds interesting in another. If that's true, then this book justifies its title, for it is essentially a meditation on a phenomenon which has obsessed me for years.

I wrote it for various reasons. Like E. M. Forster, I am one of those people who doesn't know what he thinks until he has written it down, and I wanted to sort out my ideas about what the Internet is, and what it might mean for society. Secondly, I was puzzled by – and suspicious of – some of the myths about the Net which are regularly regurgitated by the mass media: I wanted to find out *for myself* where the thing came from – who built it, and why, and how? And most of all perhaps, I wanted to express my sense of awe at what the creators of the Internet have wrought, and to try and communicate some of this wonderment to others.

Accordingly, this is an intensely personal work which makes no claims to provide a definitive account of the Net's evolution. Rather it picks out aspects of the story which seem to me to be significant, and tries to explain why. And if it reads like a passionate work, then that is because I feel passionately about its subject. The Net provides some of the things I longed for when I was young – access to information and knowledge, communication with faraway places, news from other cultures, windows into other worlds. As a boy I haunted the shelves of the local Carnegie Library, reading everything I could lay my hands on. But the more I read the more I became aware of how limited were the resources of that admirable institution.

The library had several books about the American Civil War, for example; but imagine, I thought, what it would be like to visit the Library of Congress and *see* the letters that Abraham Lincoln wrote during that terrible conflict. The encyclopaedia entries on Leonardo da Vinci were fascinating, but lacked what I really lusted to see – the notebooks written in his famous 'mirror' handwriting. A biography of Mozart merely fuelled a desire to visit the British Museum and inspect one of his manuscripts for myself, if only to verify that it was as free of corrections and revisions as the author had claimed. And a book about the history of New York prompted a yearning to see the Brooklyn Bridge and Manhattan by night.

To a child in my position in the 1950s, these were aspirations on a par with coveting a vacation on Mars. Yet today they are virtually achievable by anyone with an Internet connection. My children, who have grown up with the Net, take such wonders in their stride. They use the network for homework and leisure with the same ease as they use a dictionary or a book or the telephone. The Net has become just a pleasurable, useful background to their young lives. But while their familiarity with its riches gives their father endless pleasure, it also triggers in him the fear that has lurked in the breast of every upwardly mobile parent in history – *that one's children have no idea how lucky they are.*

And, in a way, why should they? Isn't their calm acceptance of a technological marvel symptomatic of the human condition? We take progress for granted and move on to the next Big Thing. We see further, as Newton said, and achieve more, because we stand on the shoulders of giants. In the case of the Net, however, these giants were engineers and as such are largely invisible and unsung. Why a civilisation which has been built by engineers should accord its creators such a low status is beyond me. We have no hesitation in honouring actors, musicians, artists, writers, athletes, doctors and, on occasion, scientists. But *engineers* . . . ah, that's different. They're just the guys you call out when something breaks down.

I am myself an engineer and so take this slight personally. My profession's cultural invisibility may have something to do with the fact that we are predominantly problem-solvers. Since a 'problem' is a discrepancy between a current state (A) and a desired one (B), engineers are mostly concerned with finding ways of getting from A to B. We therefore tend to focus on technical means, rather than on ends. And generally the ends are decided by someone else – a superior, a client, a government agency, a company.

Ends are chic, cool, interesting. They are the embodiment of values and aspirations. Means, on the other hand, are mundane, uncool, routine. And the engineers whose job it is to devise them tend to be unobtrusive, pragmatic, utilitarian, intellectually eclectic. They have neither the time nor the inclination for self-advertisement. They are, in Lord Beaverbrook's famous phrase, 'the boys in the back room . . . the men who do the work'. They just want to get the job done and will use any tool that looks useful, even if doing so causes purists (especially of the mathematical variety) to faint in distaste.

As a result, engineers have generally had an indifferent press. This little book is an attempt to redress the imbalance, to pay homage to the amazing ingenuity of the people who created the Internet, to rescue them from the obscurity to which our unthinking complacency generally consigns them.

We owe a debt to these architects of our future, not just because they created one of the most remarkable things humans have ever built, but because in the process they invented institutions and traditions from which society might usefully learn a thing or two. The Internet, for example, is a totally open system. There is no gatekeeper, no membership committee, no waiting list. Anyone can join in, so long as they obey the rules embedded in its technical protocols. These protocols and the software which implements them – and thereby governs the network's operation – are transparent and public. They were created by engineers and computer scientists working in a spirit of enthusiastic co-operation, debugged in the crucible of intensive peer-review and tested to the limit by the exponential growth which they have enabled. At a technical level, there are no secrets on the Net, and the result is a system of unparalleled resilience and dependability. There is – as we shall see – a direct causal connection between these two facts. Isn't there a lesson here for society about the importance of open institutions and the power of a free market in ideas?

The other interesting thing is that not a single line of the computer code which underpins the Net is proprietary; and nobody who contributed to its development has ever made a cent from intellectual property rights in it. In the age of the bottom line, where the default assumption is that people are motivated by money rather than altruism, and where software is the hottest intellectual property there is, this is surely extraordinary. The moguls who dominate the computer business would have us believe that the Internet must be an aberration – that it is the exception that proves the rule about the primacy of greed. If I've learned anything from writing this brief history it is that they've misjudged the Net. But then, they always have.

Acknowledgements

I gnorance, wrote Lytton Strachey in *Eminent Victorians*, is the first requisite of the historian – 'ignorance, which simplifies and clarifies, which selects and omits, with a placid perfection unattainable by the highest art'. In that respect at least I was eminently qualified to embark upon this book. I am grateful to the many friends and colleagues who laboured mightily to save me from myself, and wish I could blame them for the errors, omissions and solecisms which remain. But, alas, custom and practice (not to mention good manners) exclude this option, and require me instead to acknowledge my intellectual debts.

Among my Open University colleagues I am particularly grateful to Martin Weller, who shares my interest in the history of our subject, provided much encouragement, commented on several drafts and generously tolerated the book's intrusion into our joint teaching work. Ray Ison risked the wrath of his family by taking an early draft on a beach holiday and came back with criticisms so acute that they made me wince. John Monk tried (with limited success) to save me from what he once wryly called my 'Boys' Own View of History'. Scott Forrest read the manuscript with a sympathetic eye and a deep understanding of the Net. Diana Laurillard tried to rescue me from sentimentality. Dick Morris, Jake Chapman, Mary McVay and Rita Daggett each went through a draft with the care that makes working with them such an unalloyed pleasure. Richard Austin, Joel Greenberg, Roger Moore, Chris Bissell, Chris

Dillon, Karen Kear and Nick Moss answered my naive questions and made helpful suggestions on topics as diverse as the functioning of a TCP/IP stack and the processing power of a PDP-7. Gary Alexander reinforced my thinking about the significance of the Open Source movement, and Andy Reilly taught me useful lessons about the exposition of technical material.

I also owe considerable debts to my colleagues at Cambridge. Roger Needham, one of the great computer scientists of our time, encouraged me to persevere with the project and put me in touch with some of the key figures in the story. Bobbie Wells listened to an early version of the first chapter and pronounced it the kind of book people might like to read. Quentin Stafford-Fraser, the man who first put the Computer Lab's coffee-pot on the Net (and thereby triggered the Webcam avalanche), cheerfully and diligently read a couple of drafts and came back for more. Simon Moore tactfully exposed some serious gaps in my knowledge, gave me my first Linux installation and finally opened my eyes to the power of Open Source. Bill Kirkman, the ultimate Intelligent Layman, read an early draft and made many helpful comments. Jack Shepherd, the author of a bestseller or two, relentlessly bullied me into clarifying the purpose, nature and tone of the book. Lewis Tiffany, the Computer Lab's Librarian, found me documents which are to the history of networking what the Dead Sea Scrolls are to biblical scholarship. And Shon Ffowcs-Williams, who is now Master of my old college and a distinguished engineer, read a draft and encouraged me to think that other engineers might read it too.

Elsewhere in the academic world, I would like to thank Albert Lindemann for reminding me of the Californian aspects of the Net story, and Norman Gowar for inviting me to give the 1998 Science Lecture at Royal Holloway and Bedford New College, which was really just an excuse to try out the content of the book on a wider audience. Nothing, I found, concentrates the mind on a subject like the prospect of having to lecture upon it in the evening. I owe a huge debt to Gerard de Vries of the University of Amsterdam, the man who many years ago awakened my interest in the history and philosophy of technology and has tried to keep me on the scholarly rails ever since.

Among those who figure in the story, I am especially grateful to Bob Taylor, Donald Davies, Larry Roberts, Roger Scantlebury, Tim Berners-Lee, John Shoch and Mike Sendall.

As someone who has always worked as a journalist as well as an academic, I am particularly grateful to my editors at the *Observer*, Tim Adams and Jane Ferguson, for allowing me to write about the Net in a part of the paper which is not roped off with signs saying 'For Techies Only'. I owe Andrew Arends several drinks and an extended lunch for endless encouragement and a careful reading of a late draft. Steve McGookin freely gave of his expertise as Editor of FT.COM, the online edition of the *Financial Times*. Ken McGoogan read a draft and made many helpful suggestions. Gillian Reynolds kept me going with thoughtful and sometimes hilarious e-mail. Tony Holden supplied quotations, title-suggestions, gossip and general good cheer. Brian MacArthur encouraged me to read more widely by making me review books about Cyberspace for *The Times*. My agent, Jacqueline Korn, waited an unconscionable number of years for me to produce a book, and graciously affected not to be astonished when I eventually did. My editor, Toby Mundy, displayed an enthusiasm for the manuscript which went well beyond the call of duty. And Peter James edited it with remarkable tact, given the provocations of the original text.

Nothing that is read with pleasure, observed Dr Johnson, was written without pain. The trouble with books is that the pain is often experienced most acutely by others. That is why my greatest debt is to my family – to my children, who have tolerated the millstone known as 'Daddy's Book' with such forbearance, and to Sue, without whom none of it would be worth while.

Cambridge
23 March 1999

Part I
Wonderment

1:
Radio days

1997

A book-lined study, late at night. The house, to which the study is an extension, is quiet. Everyone except for a middle-aged man is asleep. He is sitting at a desk, peering at a computer screen, occasionally moving a mouse which runs on a mat on the desktop and clicking the button which sits on the back of this electronic rodent.

On the screen a picture appears, built up in stages, line by line. First it looks like a set of coloured bars. Then some more are added, and gradually the image becomes sharper and clearer until a colour picture appears in the centre of the screen. It is a cityscape. In the foreground are honey-coloured buildings and a solitary high-rise block; in the distance is the sea. Away to the left is a suspension bridge, wreathed in fog. It is in fact a panoramic view of San Francisco, snapped by a camera fixed on the roof of the Fairmont Hotel on Nob Hill. The bridge is the Golden Gate. Underneath the photograph is some text explaining that it was taken three minutes ago and will be updated every five minutes. The man sits there patiently and waits, and in a few minutes the image flickers briefly and is indeed rebuilt before his eyes. Nothing much has changed, except that the camera has moved slightly. It has begun its slow pan rightwards, towards the Bay Bridge.

And as the picture builds the solitary man smiles quietly, for to him this is a kind of miracle.

1956

Another room, another time. No book-lined study this, but a spartan bedroom. There are two single beds, covered in blankets and cheap eiderdowns. The floor is covered in linoleum, the window by a cheap unlined curtain with a floral pattern. By the wall stands a chest of drawers. The only decoration is a picture of the Sacred Heart. No books of any description can be seen.

It's late. Somewhere, in another part of the house, a man can be heard snoring. Between the beds is a table, on which sits a large radio with a brown Bakelite shell, an illuminated dial, a large tuning knob. The radio is turned towards one of the beds, in which lies a young boy who is staring intently at the dial and slowly turning the knob. The light from the dial illuminates his face. Through the ventilation slots in the back of the radio can be seen the glow of thermionic valves.

The boy is entirely focused on the sounds coming from the loudspeaker of the radio. Mostly, these are strange whistling noises, howls, bursts of static, occasional frantic sequences of Morse code. He continues gingerly to rotate the knob and eventually stops when a distant voice, speaking in English, suddenly issues from the radio's grille. The voice waxes and wanes and echoes oddly. It sounds like someone speaking through a very long pipe which is continually flexing. But the boy is transfixed by it for this is what he has been searching for – a voice from another continent, another world. He smiles quietly, for to him this is a kind of miracle.

I am that man, and was once that boy. What connects the two is a lifelong fascination – call it obsession if you like – with communication, with being able to make links to other places, other cultures, other worlds. The roots of this obsession have often puzzled me. I am not – never have been – a gregarious person. Quite the opposite. I was a solitary child and my classmates at school and university always thought of me as a loner. I was never enamoured of the noisy solidarity of pub or club. So why was I possessed of a desire to make contact with distant places?

It can partly be explained by the start I had in life. I grew up on what seemed at the time like the edge of the world – in a remote part of rural Ireland, in a household with few books, magazines or television. The only newspapers which ever crossed the threshold were the *Irish Press*, the organ of Eamonn de Valera's Fianna Fail party, and the *Irish Catholic*, an equally pious propaganda sheet for the Vatican. The only books were a few *Reader's Digest* 'condensed' books plus an edition of *Newnes Pictorial Encyclopaedia* bound in a red leatherette material which matched the upholstery of the household's two armchairs. (Only the rich and sophisticated had three-piece suites in those days.) Films were a no-go area; although there was a cinema in the town, it screened only what the parish priest approved, and even then was off-limits to us, because my mother regarded Hollywood as an agent of Satan. Our household was thus what Al Gore would call an 'information-poor' environment.

It's hard now to imagine what life was like then.[1] Foreign travel was unheard of, for example. Apart from those who emigrated to Britain or the United States, virtually nobody we knew had ever been abroad, and those who had were invariably people who had made a pilgrimage to Lourdes. Nobody in our circle ever went overseas on holiday, and no foreign languages were taught in the schools I attended – with the exception of Latin, no doubt because it was useful for those who went on to the priesthood. We lived in a closed society which thought of itself as self-sufficient, the embodiment of the ancient Republican slogan, *Sinn Fein* – Irish for 'ourselves alone'.

There was, however, one chink of light in the suffocating gloom – the radio (which we called the *wireless*; 'radio' was a posh word, unthinkable in our circle, uttered only by those who had *lunch* at the time we had *dinner*). It was by modern standards a huge apparatus powered by valves – which is why it took some time to warm up – and a 'magic eye' tuning indicator – a greenish glass circle which winked at you as the signal waxed or waned.

The tuning knob was a marvellous affair. The spindle went

through a hole in the glass of the dial. On the other side, it drove a large metal wheel to which a cord was attached. As the wheel turned, a metal bar hanging from the cord moved horizontally to indicate the wavelength[2] of the signal currently being received. A small lamp provided a backlight and, emerging through the ventilation slots at the back of the device, threw weird shadows on my bedroom wall.

The legends etched on the dial face were marvellous. There were some numerical markings, but generally stations were identified by the location of the transmitter. Thus the BBC Long Wave programme (source of *Radio Newsreel* and *Dan Dare*) was 'Droitwich'; Netherlands Radio was 'Hilversum'; Radio Luxembourg was, reasonably enough, 'Luxembourg' and Radio Eireann, our native broadcaster, was 'Athlone'. On the long-wave band I remember Minsk and Kalundbourg, Ankara on 1,700 metres and Brasov on 1,950. On medium wave there were stations based in Bordeaux, Stockholm, Rennes, Vienna and somewhere called Mühlacker. To my impressionable mind, these locations took on ludicrously glamorous connotations. I can still remember the day, decades later, when I stumbled on Droitwich on the A5 and discovered what a mundane place it really was. But back in the 1950s it seemed as exciting to me as Paris or Berlin or Istanbul.

The best thing about our wireless, though, was that it had a short-wave band. This was a source of endless fascination to me, because it meant that even with this primitive device one could listen to the world.

At first I couldn't understand how it worked. Why was reception so much better at night? Why it was so infuriatingly variable? I asked my father, who looked evasive and said it was something to do with 'the whachamacallit sphere' (he always called complicated things the whachamacallit), but this gave me enough of a steer to go to the local Carnegie Library and start digging. In due course I discovered that he was referring to the ionosphere – a layer of charged particles high up at the edge of the earth's atmosphere which acts as a kind of reflector for radio waves of certain frequencies. The reason short-wave radio could travel such vast distances was that it used the ionosphere to bounce signals round the world – which was why radio hams[3] in Latin America or Australia could sometimes be heard by a young boy on the western seaboard of Ireland. Signals from such distant shores were more likely to get through at night because then the ionosphere was higher[4] and transmission over longer distances was possible.

I was spellbound by this discovery of how technology could piggy-back on a natural phenomenon to propel low-power signals through immense distances. But most of all I was entranced by the idea of short-wave radio. For this was a technology which belonged not to great corporations or governments, but to *people*. It was possible, my father explained, to obtain a licence to operate your own short-wave radio station. And all over the globe people held such licences which enabled them to sit in their back rooms and broadcast to the whole world. And to me.

My father was simultaneously pleased and threatened by my interest in short-wave radio. It was something he himself had been passionately interested in as a young man. But even then I sensed that he also felt undermined by my youthful enthusiasm. I remember the look of alarm on his face when I returned triumphantly from the library having figured out the secret of the ionosphere. Later – much later – I realised that my naive obsession forced him to acknowledge that he had never managed to carry his own passion to its logical conclusion and become a radio ham. He thought this made him look a failure in the eyes of his eldest son.

In fact, his failure moved me more to frustration than to pity. I was mad simply because Da's non-possession of an amateur radio operator's licence meant that I was deprived of an opportunity to participate in this magical process. Listening to short-wave transmissions was bliss; but to be able to send them would be absolute heaven.

Da's interest in radio had been awakened in the 1930s, when he was an ambitious young postal clerk in Ballina, County Mayo. He had been trained in Morse (because that was the way telegrams were transmitted in rural Ireland) and to his dying day retained a wonderful fluency in that staccato lingo. It was from him, for example, that I learned that Morse operators do not listen – as Boy

Scouts do – for the dots and dashes of each letter. Instead they listen for patterns and rhythms. He illustrated this by teaching me the rhythm of CQ – the international signal which asks 'Is anybody out there listening?' The Boy Scout version is dash–dot–dash–dash, dash–dash–dot–dash. But my father sang it as dahdeedahdah, dahdahdeedah and immediately I understood what he meant.[5]

My father had left school at sixteen and gained entry to the fledgling Irish Free State's postal service by doing a competitive examination. He was highly intelligent and quietly ambitious, but these were no guarantees of preferment in the civil service. By the standards of people from his background, he was doing well to have become a postal clerk in his twenties. His position was prized because it was secure. It was, his mother crowed, a job for life. He was working for the government. But it was poorly paid.

Early on in his time in Ballina, Da became interested in amateur radio. In those days a candidate for an operator's licence had several obstacles to overcome. First he had to pass two examinations – one in Morse, the other in the physics and mathematics of wireless propagation – stuff about *frequency* and *capacitance* and *inductance* and *Ohm's Law* and *resistances in series and parallel*. The other obstacles were less formal, but no less forbidding to a man in Da's position. The aspiring broadcaster had to have suitable premises (universally called, for some reason, a 'shack'), together with sufficient land to set up a substantial antenna, and enough money to purchase the necessary kit.

Of these, the only hurdle Da had been able to overcome was the first – his Morse was more than adequate. But his truncated schooling left him terrified of the theoretical examination. As a young bachelor living in digs (a rented room in a small family house) he could not have a suitable shack. And his wages as a clerk would not have run to any fancy equipment for, like most men of his background, he sent much of it home to his parents in Connemara.

As a result, he never managed to indulge his passion for radio and instead had to content himself with looking over the shoulders of others more affluent than himself. Often the *quid pro quo* was that he taught Morse to these rich kids. In particular, there was one chap, the heir of a prosperous Ballina merchant family, who had obtained a licence in the 1930s and with whom Da had become genuinely friendly.

My father never told me any of this, of course. But he once did something which explained everything. It happened on a summer's evening in the 1950s when I was about ten. We were back in Ballina, on holiday, and he suddenly asked me if I would like to go and see Ian Clarke's shack. Knowing that Mr Clarke was a licensed ham I clambered breathlessly into the car and we drove out of the town along the banks of the Moy, eventually turning through some large gates into a winding, tree-lined drive which swept round to reveal one of those glorious Georgian boxes which were the Irish Ascendancy's main contribution to civilisation.

I had never seen a house like this up close, let alone been inside one. It had double doors, a marbled hall two storeys high and a glorious curving staircase. The drawing room had colossal windows which seemed to stretch from floor to ceiling. One set opened on to a manicured lawn which led down to the river. There was not one, not two but *three* settees arranged in an open rectangle around a coffee table. While I stood stupefied by these surroundings, a middle-aged, affable man appeared, greeted me solemnly and got my father a drink.

After a while, he turned to me and said, 'Your Daddy tells me you're interested in radio. Would you like to see my set-up?' 'Yes, please, Mr Clarke,' I burbled, terrified in case I should make a fool of myself. 'Well, come on then.' He led the way out of the room and up that great staircase. I remember how quiet the house seemed, and how our feet made no sound on the carpet. We turned into a high-ceilinged room with a large window. There was a desk with a microphone, some books and a notepad on it. To the right of the desk was a rack of equipment garnished with lights, dials, knobs and switches.

He gave me a cursory explanation of the kit, then switched the various units on – power supply, receiver, transmitter, amplifier. One by one the lights came on. Then he sat down at the desk, made an entry in a log, twiddled a knob and spoke into the mike. 'CQ, CQ, CQ, this is . . .' and here he gave his call-sign – '. . . calling.' He repeated this a few times while I held my breath. And then faintly, but distinctly, came an answering voice. An amateur radio operator somewhere in Scandinavia. He and Mr Clarke then had a conversation, probably fairly banal, and involving the ritual exchange of information about their respective technical set-ups, but all I remember of it was that he mentioned at one point that he had 'a young friend with him who hoped one day to become a ham himself', at which point I nearly passed out.

I remember little else of that evening, save that my father and Mr Clarke talked for ages about the past and of what they used to do in their bachelor days. Later, driving home in the dark, I asked, 'Da, is Mr Clarke very rich?' He replied laconically, 'Well, he doesn't want for a bob or two anyway,' and I finally understood why my father had never obtained his licence.

Why radio? Why did my childhood obsession not lead to pen pals, language courses, foreign travel? I think the immediacy and scope of the medium were the key attractions. It put you in touch with what was happening – right *now* – on the other side of the world. And (if you had that magic licence) it put you in charge of the connection process.

Of course the humble telephone provided exactly the same facility. But to say that is to ignore the realities of my boyhood. In the world in which I grew up, the telephone was anything but humble. My family did not have one. Most families we knew didn't. Making a phone call was a big deal. You had to be quite grown up to be allowed simply to answer the damn thing. Most people would send a child 'on a message' rather than make a local call. And nobody, but nobody, in my parents' circle ever made an international call. A 'trunk call' to Dublin was about the extent of it, and then only when there was a crisis of some sort – a death or a heart attack or, occasionally, a birth.

But radio, radio – that was free, and available, and much, much more exotic. For some years – I guess between the ages of nine and fourteen – I went through the time-honoured phases of an obsession. I pored over the magazine *Wireless World*, saved up pocket money to buy resistors and capacitors and diodes and crystals and transformers and induction coils and loudspeakers from smudged catalogues of radio spares, constructed radios of varying degrees of complexity and with varying degrees of success (probably because I never really mastered the art of clean and reliable soldering). For a time I hung around the only radio dealer in town, a man with an exotic past which included serving in the merchant navy. But in the end – and unlike the true fanatic – I lost interest in the technology and returned to my first love: the content, those tinny, fading, interference-ridden voices from other worlds.

The comparison of the Net with the early days of radio works not just on a personal level. The first radio enthusiasts were much ridiculed by the chattering classes of the day. They were seen as cranks with their weird equipment – crystals, antennae, coils, cat's whiskers, horns – and laughed at because of their willingness to spend long hours waiting to hear the crackling hiss of a distant broadcast. What made them seem even more absurd in the eyes of their tormentors was the fact that there was 'nothing much on the radio anyway', certainly nothing worth listening to.

Much the same is said about today's Net enthusiasts. They are ridiculed as socially challenged nerds or 'anoraks' huddled in bedrooms surrounded by equipment and glowing screens. Their willingness to persevere in the teeth of transmission delays and the limitations of low-bandwidth telephone lines is derided by those who describe the World Wide Web as the 'World Wide Wait'. The notion that any adult person would sit patiently while a low-resolution colour picture slowly chugs its way through the Net seems absurd to those who are far too busy with important matters like watching television or reviewing tomorrow's list of engagements in their personal organisers. The 'plug-ins' by which Webheads set such store – those programs which extend the capability of browsers like Netscape to handle digital audio, compressed movies and other arcane data streams – seem as weird as the cat's whiskers of yesteryear. And, of course, there is always the refrain about there being 'nothing worth while on the Internet'.

So it is deeply ironic that one of the most evocative uses of the Net is as a medium for *radio* broadcasts. My national radio station, for example, is RTE in Ireland. Although it broadcasts in stereo from the Astra satellite, I cannot receive it in Cambridge because I do not have a satellite dish. I like to listen to the news in Irish because it keeps my knowledge of the language from atrophying. There must be quite a few like me in the 70-million-strong Irish diaspora dispersed all over the globe.

Now, through a terrific piece of software technology called RealAudio,[6] I can listen to the news in Irish no matter where I am in the world. What happens is that RTE puts the daily bulletins on a computer running special software which turns it into a Real Audio server. If I wish to listen to a bulletin, I click on the hotlink to it on a Web page and, after a short interval, the rich sound of a native speaker comes rolling out of the speakers on either side of my computer. The quality is sometimes relatively poor – about the same as AM signals. Sometimes the transmission breaks up and distorts, just as those long-distant broadcasts of my youth did. But it still seems like a small miracle.

It's funny how touching this wonder can be. Last year, my eldest son was doing a project on the poetry of Robert Frost. His task was to make a film about 'The Road Not Taken' and he dropped in to ask if I knew where he could get his hands on a recording of Frost reading the poem. My first reaction was to telephone the BBC. But then I suggested – with nothing more than a mild sense of optimism – that he try hunting for it on the Web. In a few minutes he had found a scholarly archive somewhere in the United States with recordings of Frost, and shortly after that we both sat stunned, listening to the poet's gravelly voice, distorted by compression and data corruption, but still speaking directly to our hearts. And I thought: *this is what the early days of radio must have been like.*

2:
The digital beanstalk

Each venture
Is a new beginning, a raid on the inarticulate
With shabby equipment always deteriorating
In the general mess of imprecision of feeling.

T. S. Eliot, 'East Coker', pt 5, 1940

W hat is a computer? An expensive paperweight – that's what. Mine is switched off at the moment. It sits inert – a costly amalgam of plastic, metal, silicon and glass – on the desk where I am writing this with a steam-age fountain pen. My computer can't do anything right now, save perhaps serve as a doorstop.

But switch the power on and something amazing happens. *The machine comes to life!* The disk drive begins to whir, the screen shows signs of activity, noises emerge from the speakers. Eventually 'Windows 95' appears, with lots of little icons, ready to be clicked on, ready to go. Somehow, an inert object has been transformed into a powerful machine which can solve complex equations, track huge budgets, find – in seconds – a single word in a haystack of half a million documents, build an index to a 120,000-word doctoral dissertation or do a thousand other things.

This little power-up sequence happens millions of times every day all over the planet as people sit down and commence work. And yet I guess that no more than a few of them ever ponder the miracle they have triggered by simply throwing a switch.

The modern world is full of such workaday miracles. Raymond Williams, the British cultural historian, once recalled his father describing someone as 'the kind of person who turns a switch and is not surprised when the light comes on'. For as long as I can remember, I have been exactly the opposite – perpetually astonished at the effectiveness and dependability of complicated machines. I still think it's wonderful, for example, that while walking along the north Norfolk coast I can punch in a few digits on my mobile phone and be almost instantly connected to someone driving down the coast of northern California on that glorious road between Monterey and Big Sur.

In the same way I am amazed that a small number of ones and zeroes burned into a tiny chip inside my laptop can induce it to start the hard disk and commence the search for the operating system program stored thereon, so that that vital piece of software can progressively be loaded into the memory and enable the machine to function. What's happening is that the machine is, literally, pulling itself up by its own bootstraps – which is where the term 'booting up' comes from, and why that little chip which starts the whole thing off is called a 'boot ROM'.

'The impossible we do today,' says the slogan Blu-tacked to a million office partitions, 'miracles take a little longer.' Yes, but why are we so blasé about the amazing things technology can do? Is our lack of awe or curiosity simply a consequence of its power and reliability? Is technology like human spines or limbs or eyes or ears – things we take completely for granted until they go wrong? Do we only think about the electricity grid when an October storm brings down power lines and leaves us groping in the dark? Or about our computers when the hard disk crashes and we find we've lost a year's work?

But, if booting up is a miracle, then you ain't seen nothin' yet. From the back of my laptop a cable snakes to a small box with a panel of tiny lights on its front. This is a modem. From the back of that another cable runs, this time to the nearest telephone socket. It may not look much, but it's my line to the future. And it does things that two decades ago most of us regarded as simply impossible.

The modem is really just a device for connecting my computer to the public telephone system and, through it, to other computers located elsewhere. It is, in essence, a communications channel along which signals flow. Think of a communications channel as a kind of electronic pipe. The measure of the capacity of the pipe is its *bandwidth*. A low-bandwidth channel is like a narrow-bore water pipe – you can only get a small volume of water/information through it in a given period of time. High bandwidth is the opposite – a large-diameter pipe. In the communications business, as in the water industry, high bandwidth is good.

The term bandwidth comes from the spread (or band) of frequencies that the channel can pass. Frequency is measured in cycles per second or *Hertz*. Low frequencies correspond to very deep, bass sounds; high frequencies are shrill. The human ear can handle frequencies from about 50 Hertz to around 17,000 to 18,000 Hertz, which means it has a bandwidth of about 17,000 Hertz – or 17 kiloHertz (kHz).

Now the bandwidth of a channel is determined to a large extent by its physical properties. The copper cable which makes up the pipe from my computer to the telephone exchange is a fairly low-bandwidth channel. When I was an undergraduate in the 1960s we were told that the bandwidth of telephone wire was quite low – about 4 kHz if I remember correctly – which was why you could never transmit hifi music, say, over the phone. Later on, when digital signals – that is, streams of 'bits'[1] or ones and zeroes – arrived, we calculated that the most you could force through copper wire was just over a thousand bits per second. To us students, this bandwidth limitation of copper seemed as rigid a restriction as Einstein's proof that nothing could travel faster than light. It was a fact of life, a law of nature.

How come then that my computer currently spits ones and zeroes down that copper wire and into the telephone system at a rate of 28,800 bits per second? And that it will soon do it at twice that rate? What's going on? Surely the bandwidth of copper is the same today as it was when I was an undergraduate?

Of course it is. What's happened is something that none of us anticipated in the 1960s, namely that if you put a computer at either end of a copper pipe, and program those computers to do some clever mathematical operations on the bitstream, then you could increase the effective bandwidth of the copper by remarkable amounts.

My modem is in effect a small, specialised computer. It takes the streams of bits emanating from my laptop and *compresses* them using a statistical technique which squeezes out redundant bits and transmits only those needed to convey the essence of the signal. It's a bit like taking the first sentence of this paragraph and transmitting it as 'my mdm is in efct a sml, spclized cmptr'.

At the other end another modem uses the same kind of mathematics to *decompress* (that is, reconstruct) the signal and restore it to its original, verbose glory. But, because the compressed signal is much smaller than the original, you can transmit more messages through the pipe in a given period of time, and so its effective bandwidth has been miraculously increased.

Obvious, isn't it? So why didn't we foresee this in the 1960s? The answer is simple. Compression technology requires two things: one, fast, ubiquitous, cheap and very small computers; and two, someone to supply the applied mathematics needed to achieve loss-less (or at any rate non-destructive) compression of data streams. Neither of those prerequisites was available in the 1960s. They are now.

The future of the standalone computer, said a vice president of AT&T once, is about as promising as that of the standalone telephone. It was arguably the most perceptive utterance ever made by a senior corporate executive. That wire snaking out of the back of my machine to the modem is what has changed computing beyond recognition because it has transformed the computer into a communicating device. In doing so it has given me access to a world beyond imagining.

Exploration of this world starts by double-clicking on an icon of a telephone which is displayed on my screen. Shortly after that, the modem wakes up; its lights begin to flash nervously. Then there is a burst of staccato beeps – the sound of the modem dialling a telephone number somewhere. In fact it's another modem in the premises of the Scottish company which gives me access to the Internet (for a fee). This company is called an Internet Service Provider, or ISP.

After a few moments, the call is answered with a whistle, and shortly after that there is a high-pitch, burbling rasp. To those who understand these things, this is a digital conversation between my modem and the one it has just dialled. They are engaged upon a process known in the trade as 'handshaking' – establishing one another's bonda fides, like two electronic canines sniffing one another.

After a few seconds the rasping ceases as abruptly as it had begun and something happens inside my computer. A program has been launched. The screen is filled by a window labelled 'Netscape'. This is what is known as a 'browser' – so called because it offers a way of reading (or browsing) the stuff published on the Net. There is a row of buttons marked 'Back', 'Forward', 'Home' and so on. The Home button has a corny picture of a little house. Below the buttons is a white box labelled 'location' inside which some gobbledy-gook has appeared. It reads 'http://www.prestel.co.uk/'. Below this is a further row of buttons with labels like 'Handbook', 'Net Search' and 'Directory'. Below these is a large box which is currently blank.

But, even as we watch, *this* box is beginning to fill. If you look at the modem you will see that its lights are now flashing furiously. Something is coming down the telephone line from somewhere and finding its way on to my screen.

It's data – streams of ones and zeroes which when assembled by Netscape and displayed on the screen will convey information to me. In fact it's the 'Home Page' of my ISP. The information is mainly in the form of words, though there are a few simple graphics. On the top left-hand corner of the frame are the words: 'Check out the hottest selection of Web Sites including our Site of the Month'. The last four words are underlined. If I move the mouse pointer over them it suddenly metamorphoses into the image of a hand. What this means is that the phrase 'Site of the month' is a *pointer* or a 'link' to another site, and if I click on it I will immediately be linked to the site in question.

Now here's the strange thing. That destination site could be a file in another directory on my hard disk, or on that of my ISP, or it could be a file on a computer in Tashkent or Santa Monica or a million other locations. To the browser they're all the same: it will start fetching the page from wherever it's held the moment I issue the command – which I can do simply by clicking on the link. Suddenly I have the world at my fingertips.

At the bottom of the screen there is a text box which gives a running report on what the browser is doing. I click on the link. The browser says, 'Host contacted, waiting for reply'. Then, immediately afterwards, it reports 'Transferring data from www.feed.co.uk'. The screen goes red and the home page of an on-line magazine called *Feed* appears. The whole process has taken a few seconds. I do not know where the computer on which the *Feed* page is stored is physically located but my guess is that it's somewhere in the United States. Yet I collected it more or less instantaneously with a simple movement of my index finger.

Most beginners are confused about what exactly the Internet is. When they ask me, I always suggest they think about it as analogous to the tracks and signalling in a railway network – the basic infrastructure. Just as different kinds of trains, trucks and locomotives run on a rail network, so too various kinds of traffic run on the Internet. And, just as on many rail systems, different companies provide different kinds of passenger services, so the Net has different services offering various kinds of facilities.

The service which just brought me *Feed* magazine is called the *World Wide Web*, so called because it encloses the globe in a web of billions of electronic threads. When I logged on I wove such a thread from my home near Cambridge to a computer in Scotland. But it could just as easily have gone to the other side of the world.

You don't believe me? Well, try this.

First, find a 'search engine' – one of the computers which index the hundreds of millions of Web pages currently available online. Search engines ceaselessly trawl the Net, indexing what they find. The one[2] I want to use now is called AltaVista. It's located in the research labs of Digital (a well-known computer company now owned by Compaq) in Palo Alto, California. Its Web address or 'Uniform Resource Locator' (URL for short) is 'http://www.alta-vista.digital.com', and I could get it by typing that text in the 'Location' box at the top of my browser window. But because I use AltaVista a lot, I have *bookmarked* its address on a list of regularly used sites. I click on the 'Bookmarks' item on the Netscape menu bar. A drop-down menu appears on which the words 'Search engines' appear. Click on that and a list of engines appears, including AltaVista. Click on that and . . .

In less than three seconds, AltaVista has replied, placing on my screen a box in which I can type whatever it is I wish to search for. In this case it is 'Salk Institute' – the American research institute set up by Jonas Edward Salk, the inventor of the vaccine which eliminated polio – so I type that and then click on a little image of a push-button labelled 'Submit'.

My request flashes across the world to Palo Alto. AltaVista searches through its index of 30 million Web pages held on 275,600 computers and 4 million articles from over 8,000 Internet discussion groups, and then sends me a list of the best matches. If I wish to go straight to an item on the list, I simply click on it. The whole process, from my typing 'Salt Institute' to the arrival of the search result, has taken less than ten seconds. I don't know what you call that, but I call it astonishing.

It's even more remarkable when you know what goes on under the hood. For example, every computer connected to the Internet can find every other computer on the network. How? Each machine has a unique address in the form of a set of four digits separated by full stops. Permanently connected machines (like the ones in my university lab) have permanent addresses. But what about the laptop which sits on my desk unconnected to the Net for long periods? Well, when the modem dials into my Internet Service Provider, my laptop is assigned a temporary (but still unique) address which holds good for the duration of the connection. I've just checked and the address currently assigned to it is 148.176.237.68. This is called its 'Internet Protocol' or IP number, but really it's my machine's Net address.

Another intriguing thing about the Net is that it doesn't seem to do anything directly. It transmits messages by breaking them down into small chunks called 'packets' and then passing the packets between computers which are physically located all over the place. My request for information about the Salk Institute, for example, might have gone via land-line to London or Amsterdam and then via satellite to the US East Coast and then across the continent to California. But it might just as easily have gone round the back way, via Singapore and the Pacific. It all depends on which channels were available when the various switching stations on its route were deciding where to pass it next. To the Net, one channel is almost as good as another. And since everything travels at the speed of light, distance doesn't really come into the calculation.

My request was a simple short message – the characters which make up the text-string 'Salk Institute' plus the address (IP number) of my machine and some other stuff added by Netscape – so it probably was transmitted as a single electronic packet. But AltaVista has replied with a much longer message consisting of pages and pages of links related to the Salk Institute.

This reply was not transmitted – or received – as a single message, but as a number of smaller 'packets' each 1,500 characters long. A special program on the AltaVista machine in California broke its reply into many such packets, inserted them in electronic 'envelopes' which it then stamped with AltaVista's address, the Internet address of my computer plus some other information and then sent on their way. A similar program on my computer received the packets, reassembled them into the right order, sent a request for the retransmission of any missing packets and then displayed AltaVista's reply on my screen.

Now obviously for this to work, both machines must have an agreed procedure for disassembling and reassembling messages. And indeed there is such a procedure – it's called 'Transmission Control Protocol' or TCP. Put it together with the protocol for addressing individual computers and you have TCP/IP. The conception of TCP/IP was one of the great technological break-throughs of the twentieth century because, without it, the Net as we know it would not have been possible. You could say that TCP/IP is to the wired world what DNA is to the biological one.

The Internet is thus one enormous game of pass-the-packet played by hundreds of thousands of computers, all of them speaking TCP/IP unto one another. It's what engineers call a *packet-switched* system. As each of my AltaVista packets was passed along the system, every computer which handled it scanned the destination address on the 'envelope', concluded that it was addressed to another machine and passed it on in my general direction until it eventually arrived at my computer.

This means that *each* packet may have travelled via a different route. It also means that because some routes are more congested than others, the packets may have arrived in a different order from that in which they were dispatched. But the TCP program on my computer can handle all that: it checks the packets in, uses the information on their envelopes to assemble them in the correct order, requests retransmission of any which have got lost in the post, as it were, and then passes the assembled message to Netscape for display on my screen.

This strange pass-the-packet game goes on twenty-four hours a day, 365 days a year. At any given moment, billions of packets are in transit across the Net. Most get through first time but some get lost or – more likely – become snarled up in digital traffic, and are retransmitted when the destination computer realises they are missing. It all works automatically, with not much human intervention. And given its complexity, it works staggeringly well. It is remarkably resilient and apparently capable of infinite growth – to the point where some people[3] have begun to argue that the Net behaves more like an organism than a machine.

The Internet is one of the most remarkable things human beings have ever made. In terms of its impact on society, it ranks with print, the railways, the telegraph, the automobile, electric power and television. Some would equate it with print and television, the two earlier technologies which most transformed the communications environment in which people live. Yet it is potentially more powerful than both because it harnesses the intellectual leverage which print gave to mankind without being hobbled by the one-to-many nature of broadcast television.

Printing liberated people's minds by enabling the widespread dissemination of knowledge and opinion. But it remained subject to censorship and worse. Television shrank the planet into Marshall McLuhan's 'global village', but the representations it diffused were tightly controlled by editors, corporations, advertisers and governments.

The Net, in contrast, provides the first totally unrestricted, totally uncensored communication system – *ever*. It is the living embodiment of an open market in ideas. Its patron saint (if that is not a blasphemous thought for such a secular hero) is Thomas Paine, the great eighteenth-century libertarian who advanced the remarkable idea (remarkable for the time, anyway) that everyone had a right to speak his mind – and the even more remarkable notion that they had a right to be heard.

As an experiment, I once timed how long it took me to produce a Web page and publish it on the Net. Here are my notes:

Activity	Time (minutes)
Launch AOLPress (a free, easy-to-use word-processor for Web pages)	1
Set up page layout (background colour, fonts, tables and so on)	4
Compose and type 250 words of text on Rupert Murdoch's global multi-media empire	10
Log on to my ISP server	1.5
Transfer Web page to the server using WS-FTP (a free program for sending files across the Net)	0.5
Total	17

In other words, it took me seventeen minutes from a standing start to become a global publisher. In doing so I sought the agreement of no media conglomerate, required the services of no agent, submitted my page (however scatological, subversive or banal it might have been) to no editor or censor. And once the page was on my ISP's server – which is permanently connected to the Internet – it was available to anyone, anywhere in the world, who wanted to read it.

'Ah,' you say, 'but who will read it? Who knows where to look for it?' That is beside the point I am trying to make, which is about how easily one can post something for all the world to read. And anyway, there are simple ways of making sure that people can find your page: you can, for example, notify AltaVista and the other search engines which index the Web; design the page (especially the page title and header) to increase the likelihood that anyone seeking information about Rupert Murdoch's media empire will find it; and so on.

This ease of publication is one of the things which makes the Net so special. Its slogan should be James Joyce's 'Here Comes Everybody'. It's equally open to lunatics and geniuses, racists and liberals, flat-earthers and cosmologists, unworldly ascetics wishing to share their interpretations of scripture and priapic youths wanting to fantasise about *Baywatch* babes. And it is available from my desktop for the cost of a local telephone call.

As I write, it's been estimated that somewhere between 120 and 150 million people use the Net. (The estimates are revised – upwards – every week.) Sober men in suits from market-research companies estimated that by the turn of the century the network would have more than 300 million users. 'The Internet', says Andy Grove of Intel, the company whose processors power over 80 per cent of the world's computers,

is like a 20-foot tidal wave coming, and we are in kayaks. It's been coming across the Pacific for thousands of miles and gaining momentum, and it's going to lift you and drop you. We're just a step away from the point when every computer is connected to every other computer, at least in the U.S., Japan, and Europe. It affects everybody . . .[4]

The strange thing is that – still – relatively few people seem to understand its significance. And the higher you look up the social and political hierarchy the worse it gets. Many of the key figures in Western society – civil servants, government ministers, newspaper editors,[5] television moguls, newspaper columnists, Wall Street financiers, bishops, opinion-formers generally – seem blissfully or wilfully unaware of what this astonishing creation might mean for humanity.

Of course they regularly pay lip-service to the Net, and mouth bromides about its importance for education and so on. But most of them have not a clue whereof they speak. In his 1997 State of the Union Message, President Clinton referred to the Internet no fewer than six times.[6] He talked about the importance of 'bringing the Net into every school in America', as if it were a kind of plumbing – which was revealing because it showed how little he understood the nature of the thing. For the Net is not a pipe (nor, for that matter, a fire-hose) from which children drink, but a beanstalk up which they can climb, like Jack of the fairy tale, into other worlds. In fact, it was subsequently claimed[7] that Clinton had never used a computer in his life. If so, he was in good company: around the same time President Chirac of France was overheard in public asking an aide what a computer mouse was called![8]

3:
A terrible beauty?

With aching hands and bleeding feet
We dig and heap, lay stone on stone;
We bear the burden and the heat
Of the long day, and wish 'twere done.
Not till the hours of light return,
All we have built do we discern.

Matthew Arnold, 'Morality', 1852

President Clinton may have been ignorant about the Internet, but at least he had a proper sense of awe about it. Listening to his State of the Union Message (on the Net, naturally) suddenly brought to mind a book I had read as a teenager and never really understood. It was called *The Education of Henry Adams* and I'd found it in a second-hand bookshop I used to haunt as an adolescent. I bought a lot of books there – including texts which were on the secondary-school English syllabus and which sometimes got me into trouble because the editions I had purchased did not correspond with those approved for youthful eyes by the Irish Department of Education.

The *Education* is an American classic – the autobiography of a cultivated and civilised Bostonian, Henry Adams, who lived from 1838 to 1918. By profession a medieval historian who was fascinated by the great cathedrals of Europe, he came from one of those Boston 'Brahmin' families which trace their ancestors back to the

Puritans and which had become wealthy and influential in the New World. (Adams's grandfather *and* great-grandfather had both been President of the United States.) His autobiography is an extended, bemused reflection on his intellectual development written in the third person as if, somehow, its author was trying to be objective about the person he is supposed to be.

There's an extraordinary chapter in the *Education* entitled 'The Dynamo and the Virgin' in which Adams recalls how he spent from May to November 1900 haunting the halls of the Great Exposition in Paris, 'aching to absorb knowledge, and helpless to find it'. He recounts how he was taken in hand by his friend the astronomer, physicist and aeronautical pioneer Samuel Pierpont Langley, a man, says Adams, 'who knew what to study, and why and how'.

Langley took his protégé to the Exposition's highest-tech exhibit – the great hall of dynamos. These were huge machines which converted mechanical energy into electricity for transmission and distribution. As Adams became accustomed to the great gallery of machines, he began to think of the forty-foot leviathans as:

> a moral force, much as the early Christians felt the cross. The planet itself seemed less impressive, in its old-fashioned, deliberate, annual or daily revolution, than this huge wheel, revolving within arm's-length at some vertiginous speed, and barely murmuring – scarcely humming an audible warning to stand a hair's-breadth further for respect of power – while it would not wake the baby lying close against its frame. Before the end, one began to pray before it; inherited instinct taught the natural expression of man before silent and infinite force. Among the thousand symbols of ultimate energy, the dynamo was not so human as some, but it was the most expressive.[1]

Over days and days spent staring at the machine, Adams became obsessed with the dynamo and what he described as its 'occult mechanism' because of his total incomprehension of how it worked: 'Between the dynamo in the gallery of machines and the engine-house outside, the break of continuity amounted to abys-mal fracture of a historian's objects. No more relation could he discover between the steam and the electric current than between the Cross and the cathedral.'[2]

Poor old Adams, condemned to wonder but never to understand. He spent his life looking for stability – which is why he was fascinated by the social order which produced medieval cathedrals. What he loved about the Middle Ages was what he saw (perhaps mistakenly) as their ideological coherence – expressed in Catholicism and symbolised by the more or less universal veneration of the Virgin Mary. This explains the curious title of the autobiographical fragment we have just examined: Adams perceived an analogy between the modern dynamo and the medieval Virgin: he thought that the machine would be worshipped in the twentieth century in much the same way as the Virgin Mary was in the twelfth. And he feared the dynamo because the centrifugal forces generated by its great flywheel seemed to him a metaphor for the times. 'Things fall apart,' wrote W. B. Yeats, 'the centre cannot hold.' Looking back over the history of the twentieth century, perhaps Adams wasn't all that wide of the mark.

The dynamo was indeed worthy of his respect, for it was the machine which made our modern, electricity-based civilisation possible. The Internet is its contemporary equivalent: it is the thing which threatens and promises to transform our future and is just as awe-inspiring in its power and scope and potential. But how can one convey this to those who have not experienced it? A standard ploy is to quote the astonishing statistics of the thing. Nobody really knows how many people actually use the Net, but, as I write, the best estimates[3] suggest a user population of between 120 and 150 million – equivalent to the combined populations of several major industrial states. Other estimates are even higher. Then there is the dizzying growth of the thing. I'm not going to bore you with the numbers,[4] but it might help to relate them to those for more familiar media. It took radio thirty-seven years to build an audience of fifty million and television about fifteen years to reach the same number of viewers. But it took the World Wide Web *just over three years* to reach its first fifty million users!

Then there is all the stuff which is published on the Net, mainly in the form of Web pages. Again nobody knows for sure how much there is (or was, because since you started on this page a sizeable number have been added), but when Digital engineers were designing the AltaVista indexing system in 1995 they estimated that the Web then contained just under one terabyte of data – that is, just under a thousand gigabytes or a million megabytes. Just to give you a feeling for what that represents, the text of this book occupies about 400,000 bytes (0.4 megabytes), so that the estimated size of the Web in 1995 was equivalent to 2,500,000 texts of this length. 'Imagine', wrote AltaVista's creators,

> that you set out with a browser and clicked to a site and jotted down what you saw and checked to see if you had seen it before. Then you clicked on another link and another link, each time checking to see if it was new. It would take time to get to each page and more time to get links to sites you hadn't been to before. And if you built a program that went through those same operations, nonstop, twenty-four hours a day, it might only get about a page a minute, or fewer than 1500 pages a day. At that rate, it would take such a program more than 182 years to look at the 100 million pages in our index today.[5]

Several years on, God knows how much larger the resource has become. One estimate put the number of documents available on the Net in 1998 at 400 million and predicted that this would rise to 800 million by the year 2000.[6]

The trouble with these stupefying statistics is that they do not actually convey the true scale of the phenomenon. What does it *mean* to have ninety – or 190 or 250 – million people online? Who can imagine such a thing? Immanuel Kant once observed that the capacity to 'take in' great magnitudes is ultimately aesthetic rather than purely logical. He illustrated this with a French general's account of how, on a visit to the great Pyramids, he was unsure about how to truly *feel* the emotional impact of their sheer enormity. Approach too close and you see only stone upon stone and miss the full sweep from base to apex; move too far away, on the other hand, and you lose the sense of awe that something so vast was constructed block by back-breaking block. (The biggest stones, remember, weighed upwards of 200 tons and were up to nine yards long, and many of them were hauled up a ramp which must have been at least a mile long.)

However the data are discounted and adjusted and filtered, though, the inescapable fact remains that the Internet has been growing at near-exponential rates for years and shows no signs of slowing down. We have therefore long passed the point where it was sensible to conjecture that the thing might turn out to be the CB radio *de nos jours*: this is no fad or passing fancy, but a fundamental shift in our communications environment. This is for real. The Net is where it's @, to use the symbology of the digital age. We have hitched a ride on a rocket, and none of us has a clue where it's headed.

There is a strange dichotomy in people's attitude to the Internet. Those who use it are, in general, pretty enthusiastic. They may moan about its disadvantages – in particular about how frustrating it is to wait for Web pages to chug their way across the Net – but in the main they value what it provides and appreciate its potential.

Everyone else seems either hostile to, fearful about or ignorant of it. As I write there's an ad for IBM on television. Two men are sitting at a table. One is reading a newspaper, the other is typing desultorily on a laptop computer. 'It says here,' says the reader, 'the Internet is the future of business. (Pause.) We have to be on the Internet.'

'Why?' asks his companion.

'It doesn't say.'

I wrote a column about television for nearly a decade in the *Observer*, a mainstream British Sunday newspaper, and had a large contingent of vociferous and opinionated readers who often wrote in to disagree (and occasionally agree) with my judgements about programmes and issues. I tried to write about television not as a separate artistic form but as a part of life – as a mirror on society – and my readers responded in kind.

Now I write a column about the Internet for the same newspaper and in much the same spirit. I still get lots of letters and e-mail, but not from former readers of my TV column. Many of those readers tell me they feel unable to engage with my new material. It is as if

the spirit of Henry Adams is alive and well and living in contemporary Britain. 'I'm sorry,' writes a woman who was once the
literary editor of a leading British magazine, 'but I'm afraid this
Internet stuff is quite beyond me.' Others tell me that the Net is 'not
for us'. It is as if I have suddenly begun to write dispatches from
some alien civilisation. In a way, I suppose I have, because most
people get their opinions about the Internet from the mainstream
media. And the impression these media create is warped beyond
caricature.

For decades, the Net was below the radar of mainstream journalism – its practitioners didn't even know of its existence. When I
used to tell newspaper colleagues that I'd been using e-mail from
home since 1975, they would make placatory noises and check the
route to the nearest exit. Some time around 1995, that began to
change – the conventional media realised something was up. But
what? At first the Net was treated as a kind of low-status weirdo
craze, akin perhaps to CB radio and the use of metal detectors. Then
it became the agent of Satan, a conduit for pornography, political
extremism and subversion. Next it was the Great White Hope of
Western capitalism, a magical combination of shopping mall and
superhighway which would enable us to make purchases without
leaving our recliners. Then it was a cosmic failure because it turned
out that Internet commerce was – shock! horror! – *insecure*. And so it
went, and goes, on, an endless recycling of myths from newspaper
clippings, ignorance, prejudice, fear, intellectual sloth and plain,
ornery malice.

Myths about the evils of the Internet are invariably leavened with
some hoary staples. The first concerns children – and the dangers
that lie in wait for them in Cyberspace. Yet the minute one begins
to probe into this phobia, strange contradictions appear. For
example, I have met parents who seem relaxed about violent videos
and/or unpleasant computer games, and yet are terrified of letting
their kids on to the Net. And when I suggest putting the PC in the
living room where they can keep an eye on it they look at me as if I
have just discovered gravity.

Another staple is the supposed threat of 'Internet addiction'. I've
just seen a newspaper story headlined 'A lot of knowledge is a

dangerous thing for addicts of the Internet'.[7] There is a large photograph of a thirteen-year-old boy eating a sandwich by his computer, plus the information that he spends 'up to four hours a day on the Internet': 'Information is becoming the drug of the Nineties, according to a research report that has found that more than half of managers and schoolchildren crave information and experience a "high" when they find the right nugget of knowledge.' Wow! People seek information *and* feel elated when they find it! And doing so from their own desks without having to trudge to a library, seek guidance from an overworked staff member and then discover that the current edition of the reference work they seek is not stocked, probably because of budget cuts. Whatever next!

To anyone with a long memory, there is a rich irony in the hostility of print journalists towards the Net because they are reacting to it much as their 1950s counterparts reacted to the arrival of television news. Then, as now, the first instinct was to rubbish and undermine the intruder. Television news, said the newspapers, would be vulgar and sensational. There would be fewer words in an entire thirty-minute news bulletin than in half a page of *The Times*. And so on.

But when denigration failed a more radical reappraisal of the role of newspapers had to be undertaken. Once it became clear that most people were getting their 'hot' news from the new medium, newspapers were forced to reinvent themselves. The broadsheets opted for providing more background, features, comment and interpretation. The tabloids, for their part, fastened like leeches on the showbiz side of television.

Something similar happened with the Net. Print journalists eventually began to realise that it might turn out to be a serious rival to their existing medium. Their attempts to ridicule the upstart had palpably failed. Instead the thing was spreading through society like a virus. As a strategy, denial simply hadn't worked. So they escalated the attack to another level and sought to undermine the upstart's credibility.

'Never mind the quantity,' was the new slogan, 'where's the quality?' The fact that there is a super-abundance of information on the Net means (we are told) that the consumer doesn't know who to trust. A Web page written by a flat-earther may look as plausible as one from the Institute of Fiscal Policy. What looks like objective product reviews may really be concealed propaganda from Microsoft (or, for that matter, from Sun Microsystems, Microsoft's arch enemy). And so on. In such a glutted environment (so the story goes) it's best to stick with sources which have established reputations for editorial independence, journalistic competence and so on – that is, *us*.

Ho, ho! The idea that newspapers are paragons of editorial objectivity will raise a hollow laugh from anyone who has been reported by them. Freedom of the (printed) press often means the freedom to publish as much of conventional wisdom and the proprietor's prejudices as will fit between the advertisements. Journalists who live in glass houses should undress in the dark.

One of the more interesting aspects of media hypocrisy about the Net is that it is not confined to the tabloid end of the spectrum. On the contrary: the worst offenders are often to be found in those publications which claim a freehold on the ethical high ground of journalism. When I was researching this chapter I came across some startling lapses in journalistic standards perpetrated by *Time* magazine, CNN, the *Washington Post*, the *New York Times*, the *New Yorker*, the *Los Angeles Times* and the *Financial Times* – all of which are pretty high-grade outfits.

In the summer of 1995, for example, *Time* ran a cover story about the menace of 'Cyberporn' which claimed that the Internet was dominated by pornography.[8] The story was largely based on a research study, *Marketing Pornography on the Information Superhighway*, carried out at prestigious Carnegie-Mellon University by someone named Marty Rimm. The report was immediately challenged by a number of academic experts on the grounds that Rimm's research methodology was deeply flawed and likely to lead to misleading conclusions. He had, for example, grossly exaggerated the extent of pornography on the Internet by conflating findings from private adult-bulletin-board systems which require credit cards for payments (and are off limits to minors) with those from the public networks (which are not).

Critics also pointed out that many of Rimm's statistics – for example, his claim that 83.5 per cent of the images stored on the Usenet newsgroups are pornographic – were nonsensical. Two acknowledged experts from Vanderbilt University, Donna Hoffman and Thomas Novak, maintained that pornographic files represent less than one half of 1 per cent of all messages posted on the Internet. It was also pointed out that it is impossible to count the number of times those files are downloaded since the Net measures only how many people are presented with the opportunity to download, not how many actually do.

This was bad enough, but there was more to come. It turned out that this learned research study was written by an *undergraduate* who already had a track record of hi-jacking public debate. At the age of sixteen, for example, while still a junior at Atlantic City High School, Rimm had conducted a survey that purported to show that 64 per cent of his school's students had illicitly gambled at the city's casinos. Widely publicised (and strongly criticised by the casinos as inaccurate), the survey inspired the New Jersey legislature to raise the gambling age in casinos from eighteen to twenty-one.

More interesting still was the revelation that Rimm was the accomplished author of two privately published books. The first, a salacious novel entitled *An American Playground*, was based on his experiences with casinos. The other, also privately published, was *The Pornographer's Handbook: How to Exploit Women, Dupe Men & Make Lots of Money*. When challenged about it, Rimm claimed it was a satire; but others perceived it as offering practical advice to adult-bulletin-board operators about how to market pornographic images effectively.

In the end, everyone involved – including Carnegie-Mellon University – backed away from the ludicrous 'study' which had fuelled *Time*'s great exposé of pornography on the Net. Four weeks after its publication, the magazine itself ran a story explaining what had gone wrong.[9] But did it apologise to its readers for misleading them? Did it hell. Attacks on the Net apparently bring mainstream media into a fact-free zone.

But since it's come up – and because it looms large in some people's prejudices about the Net – let's tackle the question of pornography head-on. What is the fuss really about? Only a fool would deny that there's a lot of porn on the Net. If you doubt it type 'cunt' (or any other relevant term that comes to mind) into a search engine and watch as it dredges up thousands, perhaps tens of thousands, of links. Click on any of them and you will be taken to a variety of sites promising glimpses of all kinds of depravity and exotica. Often there will be an explicit image or two to tempt you further. After that, in many cases, the fun stops because to get at the really hard stuff you have to pay – usually by credit card. Most pornographers, after all, are not charities. They did not get where they are today by giving things away.

The credit-card hurdle is where most minors seeking access to online porn would fall. But there are still a lot of 'free' porn sites, offering photographs of varying degrees of explicitness, which, like most parents, I wouldn't want my children to see. Even so, I have deliberately not installed on my household's machines the so-called 'censorware' or filtering software often touted as the solution to the anxious parent's terror of online filth. Instead I keep the PCs downstairs where I can see them – and where I can monitor what the children do with them, rather than banishing them to their bedrooms. The solution to the problem, in other words, is to exercise one's parental responsibilities, not subcontract them to software written by someone else.

But although most of the fuss in the media is about the threat to children posed by online pornography, that's just a smoke-screen for the real issue – which is adult sexuality. Given that the Net is the first completely free publication medium that has ever existed, it would be astonishing if there were *not* lots of porno-graphy on it because the human appetite for porn is – and always has been – vast and unimaginably diverse. And the history of communications technologies suggest that pornographers have always been early and ingenious adopters of the latest develop-ments.

What makes the Net unusual is that it is the first conduit for illicit or disreputable publications which does not require the consumer to cross a shame threshold. If you want to buy a raunchy magazine or a smutty video, you have to run some risks of exposure or embarrassment. You may have to visit a sex-shop, for example, and be seen emerging from the premises by a colleague; or receive a parcel in the mail that could be opened in error by someone in your family or office; or undergo the humiliation of having to ask the newsagent to take down a particular magazine from the top rack where it is displayed beyond the reach of children and the purview of respectable citizens; and when you get it home there is always the problem of where to hide it.

But with the Net – well, you simply pay by credit card and, in the privacy of your own study or living room, get whatever turns you on. *And nobody – spouse, partner, family, colleague or friend – need ever know!* So it seems to me that the really interesting thing about online porn is not that the Net carries it, but that apparently large numbers of people use it for precisely this purpose. And *that* tells us much more about human nature than it does about technology. It may also explain why people get so steamed up about Internet porn: the phenomenon reveals some unpalatable truths about ourselves.

Yet nobody talks about this. The person who first opened my eyes to its significance is Robert M. Young, a psychotherapist and philosopher of science who has a good understanding of both the Net and the human psyche and has dared to write about both. In a thoughtful article on 'Sexuality and the Internet'[10] he describes a patient of his who uses the Web for sexual release. The patient, Young writes,

> spends some time every night on the internet while his partner is asleep. Until recently he spent up to two hours a day searching for sites concerned with spanking. He'd enter that word on a search engine and then work his way through the sites which came up. He eventually found one which was exquisitely suited to his tastes and can now gratify his daily needs in about half an hour.

This man shares his flat with a woman friend with whom he professes to want to have a long-term relationship, but, to tell the truth, he prefers to masturbate to the pictures on the spanking Websites rather than make love to her. In fact, he tells Young that the first time he is with a woman he is excited by her, the second time bored, and by the third he despises her.

He used to use spanking magazines and videos and prostitutes specialising in spanking, but now he prefers the pictures on the internet. It is private, instantly accessible, and there is no real human interaction – no relationship – to conduct except with a fantasy aided by pictures of strangers with exposed bottoms. The women in the internet pictures are always compliant and never answer back. Needless to say, my patient is reluctant to acknowledge that his preoccupation with spanking is vengeful or retaliatory.

Although he does not push the argument this far, what I infer from Young's analysis is that the apparent popularity of online pornography is really a commentary on the difficulties people have in engaging in real relationships with real partners. Internet sex, Young writes, 'is the fantasy alternative to real intimacy'. He's right: it's a defence against real relationships. And the reason people rant about the Net in this context is that it has begun to uncover the extent of the problem. No wonder they want to shoot the bearer of such unpalatable news.

Asking whether the Net is a good or a bad thing is a waste of time. People once asked the same rhetorical questions about electricity and the telephone. A much more interesting questions is this: *What* is the Net? The conventional answer (the one you find in dictionaries) is that it's a 'global network of computer networks', which implies that the Internet is some kind of global machine. Now a machine is something that can be switched off. Could you switch off the Net? Answer: only in principle. It would involve shutting down a large proportion of the most powerful computers on the planet, and making sure they did not reboot automatically. In fact, as we will see later, one of the neat side-benefits of the Internet's original design was that it could continue to pass messages even if large chunks of it were irretrievably damaged by some catastrophe like an earthquake or a nuclear exchange. Under such dire circumstances, it might hiccup a bit, and messages might take much longer to reach their destinations, but they would get through in the end. So if the thing is resilient enough to ride out a nuclear strike, we are forced to conclude that, in practical terms, it cannot be switched off.

The Net is therefore not a machine, but it can function as one, and when it does it's astonishingly powerful. In 1997, for example, it was used by a group of researchers as a giant supercomputer to unscramble a message which had been coded with heavy-duty encryption software. The message was thought to be very secure because in order to break the code you would have to try 281 trillion possible keys to be sure of finding the one that fits. It was estimated that the fastest desktop computer available at the time would take about two years working round the clock to do the job, yet only thirteen days after the coded message was released a German programmer successfully deciphered it. If he hadn't cracked it, then a Swede or a South African or any one of more than 5,000 volunteers, linked via the Net, would have done.

In this fascinating experiment, the volunteers had divided up the problem between them. All of them had computers which are permanently or regularly hooked up to the Net. Such machines spend much of their time idling – waiting for something to happen like the arrival of a request for a Web page – and the idea was to use some of this free time to tackle a common task. Each participating computer downloaded from a central co-ordinating Website a small program which checks a group of keys when the computer is otherwise unoccupied. On finishing its calculations, each machine posted the results back to the Website and downloaded more untested keys to try.[11] 'For two weeks,' wrote W. Wayt Gibbs,[12] 'this simple scheme created an ad hoc supercomputer as powerful as any yet built.'

And it's only the beginning. The co-operative use of networked computers is now so routine that there is a technical term – *metacomputing* – for it. It's an obvious application of Metcalfe's Law, which says that the power of a network increases as the square of the number of computers connected to it.[13] But the Net's burgeoning population of densely interconnected computers is leading some people to ask whether this is something which is not only quantitatively different from anything that has gone before, but actually different in kind. In a provocative book, *Darwin among the Machines*,[14] George Dyson argues for example that the networked computers which now surround us by the million constitute, collectively, a form of intelligent life – one that's evolving in ways we may never understand.

Dyson portrays the Net as a globally networked, electronic, sentient being. His analysis leans heavily on biological metaphors and the theory of symbiogenesis proposed in the early years of this century by a Russian botanist named Konstantin Merezhkovsky. Symbiogenesis offered a controversial complement to Darwinism by ascribing the complexity of living organisms to a succession of symbiotic associations between simpler living forms. What Dyson has done is to apply this idea to computing.

In his scheme of things, fragments of software replicating across the Internet are analogous to strings of DNA replicating in living cells. As in biology, evolutionary pressure provides the drive: what works survives; what doesn't gets deleted. Thus Dyson treats operating systems – the programs which govern the operations of a functioning computer – as 'complex numerical symbioorganisms'. The most successful such 'organisms' – the operating systems like MS-DOS and Windows and UNIX which run the types of computer that have dominated the market-place – succeeded in transforming and expanding the digital universe so as better to propagate themselves. Their evolutionary 'success' depended on the number of machines they got to run on. The success of the machines in turn depended on their ability to support the successful software, those that clung to moribund operating systems rapidly became extinct.

Dyson describes three stages in this strange kind of symbiotic evolution. The first was the creation and dissemination (via magnetic tape and disk) of the mainframe and minicomputer operating systems of the 1960s and 1970s. Then came the microprocessor, which led to the replication of processors thousands and millions at a time, and to the dissemination of their operating systems on floppy disks. 'Corporations came and went,' he writes, 'but successful code lived on' (as we discovered when corporations begin to grapple with the 'Millennium Bug', the two-digit date field planted in the database systems of the 1960s and 1970s).

The third stage came twenty years later, and was driven by the 'epidemic' of technical protocols like TCP/IP which underpin the

Internet and which propagate at the speed of light instead of at the speed of circulating floppy disks. As mainframes gave way to minicomputers and then to networked PCs these software organisms began competing not only for memory and processor time within their local hosts but within a multitude of hosts at the same moment. Successful code (UNIX, Windows 95/98) is now executed in millions of places at once, just as a successful genetic program is expressed within each of an organism's many cells.

The central thrust of Dyson's book is that the Net is qualitatively different from other technological systems like the electricity grid. He thinks it might have 'emergent' behaviours which we do not (cannot?) anticipate – that is to say, behaviours which cannot be predicted through analysis of any level simpler than that of the system as a whole. In a way, this is just an extension of the view that what we call 'intelligence' is really an emergent property of the human brain. One of the great mysteries of life is how the brain, which contains only relatively slow computing elements, can nevertheless perform cognitive feats way beyond the reach of even the most powerful supercomputers. The conjecture is that it is the brain's incredibly dense web of interconnections which gives us these powers. Although simple by comparison with the brain, the Internet is also a very densely interconnected system and maybe one day new forms of artificial intelligence will emerge from it.

In searching for a metaphor to describe the Net, Dyson goes back to Thomas Hobbes's portrayal of society as an intelligent 'Commonwealth' – that is, 'a self-organising system possessed of a life and intelligence of its own'.[15] And he draws the title of his book from an essay of the same title written in 1880 by that erratic Victorian, Samuel Butler, who wondered whether machines would 'ultimately become as complicated as we are, or at any rate complicated enough to be called living'.

The conjecture in Dyson's book – that the Net may be the primordial digital soup out of which new forms of artificial intelligence might arise – is breathtaking. Ten years ago, I would have denounced it as baloney. Today, it still seems far-fetched. Tomorrow – who knows?

And anyway, you don't need to resort to theories about the

mechanics of digital replication to explain why the Net is different from anything we've seen before. For the conventional definition describing it as a global network of computer networks contains an elementary schoolboy mistake. It makes no mention of people.

The Net is really a system which links together a vast number of computers *and the people who use them*. And it's the people who make it really interesting. They use it for lots of things that human beings like to do (including, increasingly, shopping) but mostly they use it to communicate. It's the phenomenon of hundreds of millions of people freely communicating via such an efficient and uncensored medium that gives the Net its special character, and its extraordinary power.

In global terms, the wired population, though sizeable, is only a subculture, and a peculiar one at that. It's heavily skewed towards the developed world, for example; indeed, how could it be otherwise when there are more telephones in Manhattan than in the whole of Africa? And, even within Western countries, the 'digerati' tend to be drawn from their affluent, educated, articulate elites – though that may be changing faster than many people realise.[16]

Howard Rheingold, who is the nearest thing Cyberspace has to an elder statesman, denies there is such a thing as a single, monolithic, online culture. 'It's more like an ecosystem of subcultures,' he says,

> some frivolous, others serious. The cutting edge of scientific discourse is migrating to virtual communities, where you can read the electronic pre-printed reports of molecular biologists and cognitive scientists. At the same time, activists and educational reformers are using the same medium as a political tool. You can use virtual communities to find a date, sell a lawnmower, publish a novel, conduct a meeting. Some people use virtual communities as a form of psychotherapy. Others . . . spend eighty hours a week or more pretending they are somebody else, living a life that does not exist outside a computer.[17]

Rheingold, who has been observing the Net from the beginning, thinks that people use the network in two basic ways: for entertainment and information; and to form what he calls 'virtual communities'.

To people who have never ventured online, the idea of a virtual community must seem at best oxymoronic, at worst absurd. How can you have 'communities' of people who never meet, who mostly know very little about one another, and who are not linked by the bonds of obligation and custom which govern real communities? Answer: you can't, and it is stretching the term to refer to many of the myriad groupings which characterise Cyberspace today as 'communities'. Real communities contain people you dislike or mistrust, as well as people who share your values. For that reason 'virtual communities' are often more like clubs or special-interest groups in which people gather to share their obsessions or concerns: Psion computers; ice-hockey; human rights violations in East Timor; quilting; the novels of Saul Bellow; white supremacy; Michael Jackson; M-series Leica cameras; political repression in Albania; the Glass Bead Game . . . you name it, there's probably an Internet discussion group for it.

'How far can mediated contacts constitute community?' asks Frank Weinreich, a sociologist who has done a study of German bulletin-board users.

> I believe they cannot. You may get to know other people through CMC [Computer-Mediated Communication], the Net will provide the means to maintain contact and interconnections between people and organization. But they won't constitute communities because CMC cannot substitute for the sensual experience of meeting one another face-to-face. Trust, cooperation, friendship and community are based on contacts in the sensual world. You communicate through networks but you don't live in them.[18]

But, having said all that, there are also groupings on the Net for which the term 'community' does not seem entirely inappropriate. The most famous is the WELL, a San Francisco-based online group founded in the 1980s by a number of ageing hippy entrepreneurs and visionaries originally centred on the *Whole Earth Catalog*, the 1968 handbook by Stewart Brand which became the unofficial bible of the counter-culture movement. The *Catalog* was a best seller and Brand ploughed some of his resulting earnings into starting the WELL. The name was derived – tortuously – from 'Whole Earth 'Lectronic Link'.

Marc Smith, a sociologist who has studied it, maintains that the social 'glue' which binds the WELL into a genuine community is a complex blend of participants' social skills, their amazing collective fund of expert knowledge (these are mainly Bay Area professionals, remember) and their willingness to support community members when they are in trouble.[19]

This last characteristic is something which also lifts some special-interest online groups into a different category. Everybody who has spent any time on the Net knows of the extraordinary opportunities it provides for reaching out and helping others. 'Here are a few people to talk to about the menace of Cyberspace,' fumes Rheingold, with heavy irony:

> The Alzheimer's caregiver afraid to leave the house who dials in nightly to a support group; the bright student in a one-room Saskatchewan school house researching a paper four hours from the nearest library; the suicidally depressed gay teenager; AIDS patients sharing the latest treatment information; political activists using the Net to report, persuade, inform; and the disabled, ill and elderly, whose minds are alive but who can't leave their beds. For them and for potentially millions of others like them, Cyberspace is not just a lifeline, it can be better than the offline world.[20]

The most paradoxical thing about the people who make up the Internet is that they are unimaginably diverse and yet at the same time are capable of behaving in quite concerted ways. One sees this most frequently whenever outside agencies (the US federal government, Scientology, authoritarian regimes in Asia, whatever) seek to interfere with the network or curtail its cherished freedoms. 'The Net', wrote one of its early evangelists, John Gilmore, 'interprets censorship as damage and routes around it.'

When the US Congress in 1995 passed a law (the Communications Decency Act) attempting to regulate what could be transmitted over the Net, American politicians were astonished at the effectiveness and coherence with which the Internet community responded to the threat. Over 20,000 Websites – some of them heavily trafficked sites – turned their pages black in protest. Senators and Congressmen were deluged with critical or angry e-mail. Heavy legal artillery was trained on the Act – paid for by outfits like the Electronic Frontier Foundation as well as the American Congress for Civil Liberties. In the end the Act was declared unconstitutional by the Supreme Court in June 1997 and American legislators are still wondering what hit them.

The coherence of the Net is sometimes its most surprising quality. After all, it has no central authority, no body which determines who can join it and under what conditions. Anyone can hook a computer to the Internet, provided they can pay for the physical connection and run the standard software. Indeed, that is precisely the way it grows at such a fantastic rate. The thing is incredibly complex – and yet it works astonishingly well, passing enormous volumes of data every day with remarkable reliability. It is an example of a self-organising system, of something where there is order without control.

The computer graphics expert Loren Carpenter once invented a game which gives a flavour of this. At a 1991 computer conference in Las Vegas, he gave every one of the 5,000 delegates a cardboard wand coloured red on one side and green on the other. The delegates were then gathered in an auditorium facing a huge video screen linked to a computer running the ancient computer game called 'Pong'. It's like a crude electronic version of ping-pong – a white dot bounces around inside a square and two movable rectangles act as paddles off which the ball bounces. At the back of the hall, a video-camera scanned the sea of wands and fed its images back to a bank of computers which calculated what proportion were showing red, and what proportion green.

Delegates in the left-hand side of the auditorium controlled the left-hand paddle; those on the right controlled the right-hand one. Anyone who thought the paddle should go up was told to flash the red side of their wand; if they thought it should go down they were to flash green. When everything was ready Carpenter shouted 'go' and all hell broke loose. Kevin Kelly was there:

> The audience roars in delight. Without a moment's hesitation, 5,000 are playing a reasonably good game of Pong. Each move of the paddle is the average of several thousand players' intentions. The

sensation is unnerving. The paddle usually does what you intend, but not always. When it doesn't, you find yourself spending as much attention trying to anticipate the paddle as the incoming ball. One is definitely aware of another intelligence online: it's this hollering mob.[21]

When I first heard about this experiment the hairs rose on the back of my neck because it captured the essence of my own experience of the Net. The only thing that is misleading about Kelly's description is that throwaway line at the end about the 'hollering mob'. This isn't a mob but a group of independent individuals, each endowed with free-will, engaged upon a common enterprise, which in this case happens to be a ridiculous computer game, but could be something much more important. Like preserving freedom of speech. Or protecting cultural diversity against the dumbing inanities of the global multi-media conglomerates. Or enabling people to do a million other things that human beings like to do.

So what is this thing we call the Net?

Whenever I think about it, what comes to mind is a line from Yeats's poem *Easter 1916* – *A terrible beauty is born*. It was his metaphor for the resurgence of Irish nationalism which followed an armed uprising against British rule on Easter Monday 1916. In military and ideological terms the insurrection was a fiasco: it was initially unpopular with the general public and was easily put down by the British. But the clumsy execution of the rebel leaders stirred some deep instinct in the Irish people which finally led them to throw off the colonial yoke. It was this awakening that Yeats sought to capture in his famous phrase; it expressed the perception that some powerful genie had been released and was now prowling the land.

Similarly with the Net. A force of unimaginable power – a Leviathan, to use a Biblical (and Hobbesian) phrase – is loose in our world, and we are as yet barely aware of it. It is already changing the way we communicate, work, trade, entertain and learn; soon it will transform the ways we live and earn. Perhaps one day it will even change the way we think. It will undermine established industries and create new ones. It challenges traditional notions of sovereignty, makes a mockery of national frontiers and continental barriers and ignores cultural sensitivities. It accelerates the rate of technological change to the point where even those who are supposed to be riding the crest of the wave begin to complain of 'change fatigue'.[22]

In its conjunction of the words 'terrible' and 'beauty', Yeats's phrase has a strange ambiguity. His critics focus on the noun and read into it the glorification of blood-soaked Irish nationalism, the ideology which found its modern expression in the terrorism of the IRA. But the adjective is equally plangent. It implies that there was something awe-inspiring as well as terrifying in the sleeping giant which the British had awoken.

Much the same might be said about the Net. Like all powerful technologies, it has an immense capacity for both good and evil. It gives freedom of speech its biggest boost since the US Constitution got its First Amendment; but by the same token it gives racists, paedophiles and pornographers a distribution system beyond their wildest dreams. The freedom it gives me to live and work almost anywhere is the upside of the freedom it gives employers to lay off office staff and contract out their work to tele-workers on the other side of the world. The Net enables us to create 'virtual communities' of geographically dispersed people with common interests; but the industries it supplants once supported real communities of people living in close proximity to one another.

The truth is that the Net is wonderful in what it can do for us, and terrifying in what it might do to us. Yeats got it about right: a terrible beauty *has* been born.

Part II
A brief history of the future

4:
Origins

There is a history in all men's lives,
Figuring the nature of the times deceased,
The which observed, a man may prophesy,
With a near aim, of the main chance of things
As yet not come to life, which in their seeds
And weak beginnings lie intreasurèd.

Shakespeare, *Henry IV, Part 2*, 1600

It's always earlier than you think. Whenever you go looking for the origins of any significant technological development you find that the more you learn about it, the deeper its roots seem to tunnel into the past. When my academic colleagues discovered I was writing this book, some of them[1] came to me with curious fragments – ideas they had picked up over the years and filed in that part of memory labelled 'you never know when this might come in useful'.

What these fragments suggested is that some of the ideas behind modern information technologies had long lurked in what Jung called the collective unconscious. The best-known example of this, I guess, is Arthur C. Clark's premonitions about satellite communications and 'intelligent' machines (the latter eventually finding global celebrity as HAL in Kubrick's movie *2001*), but there are other, less well-known, cases.

Consider, for example, the strange case of 'The Machine Stops', a short story which contains a vivid description of what is nowadays called 'virtual reality' but which was written before the First World War by that most untechnological of novelists, E. M. Forster.[2] It was composed, Forster wrote, as 'a counterblast to one of the heavens of H. G. Wells' but is obviously also a satire on mankind's increasing dependence on technology, and it includes references to 'cinema-taphoes' (machines which project visual images) and a facility for instantaneous simultaneous correspondence with multiple persons (e-mail?).

Then there is the curious coincidence that in 1946 Admiral Lord Louis Mountbatten made a speech saying this:

> In the field of communications it was hoped that a standard system would be evolved, in which Morse operators would be replaced by automatic apparatus such as the teleprinters and . . . facsimile transmitters, and that a single unified world-wide network of stations would be set up over which messages would be passed automatically . . . from origin to destination without appreciable delay at any necessary intermediate stations.[3]

And what about the strange fact that in a 1966 story entitled 'The Moon Is a March Mistress' the science-fiction writer Robert Heinlein described a multi-computer network operating on similar principles to the Internet, containing links to all known data and capable of communicating with people over ordinary phone lines?

How far down should we drill in seeking the origins of the Net? Given that a large part of my story is about computers, should I go back all the way to the 1830s when Charles Babbage developed detailed plans for what he called the 'analytical engine', a device capable of performing any arithmetical operation in response to instructions contained on punched cards? Babbage's engine had a memory unit for storing numbers, sequential processing and many other elements of a modern computer. And although he never completed the construction of the engine (for various reasons including his fidgety temperament and the difficulty of making precision-engineered components), we know that the design was sound because in 1991 some researchers at the Science Museum in London built a working model (accurate to thirty-one digits) to

Babbage's specification. (Five years later, Nathan Myhrvold, one of Microsoft's top honchos, gave the Museum £750,000 to build a replica of the computer plus a working model of its printer for the lobby of his Seattle mansion.)[4]

But why stop with Babbage? After all, one of the key features of his engine was its use of punched cards – an idea borrowed from the French inventor Joseph-Marie Jacquard, who in 1804 introduced an automatic weaving loom controlled by punched cards. And if Jacquard gets a look-in, why not George Boole (1815–64), who was Professor of Mathematics at my first university – University College, Cork – and invented the type of logic used by digital computers?

But then, why stop at the nineteenth century? After all, did not Kepler's friend Wilhelm Schickard invent the first mechanical calculator round about 1623? And what about Blaise Pascal, who built an adding machine – a digital device with numbers entered by dial wheels – in 1642? Or Leibniz, who in 1673 developed a more advanced calculator than Pascal's which could multiply, divide and calculate square roots? Or the Frenchman Charles Xavier Thomas de Colmar who in 1820 produced the first commercially available calculator, the arithmometer?

You can see the problem: any starting-point for an historical trail is likely to be arbitrary. So let's try a different tack.

How did life begin? Some biologists speculate that there was once a 'primordial soup' of molecules which was radically disrupted by an external event like a flash of lightning which provided the energy necessary to start things off. I have no idea whether this is plausible, but it provides a metaphor for thinking about human creativity. For no man is an island, not even the greatest of geniuses. We all swim in a soup of ideas, and occasionally lightning strikes, sparking off thought-processes which lead in unexpected – sometimes unimaginable – directions.

In looking back at the origins of the Net, the first thing we find is that the primordial soup was concentrated in a single spot on the East Coast of the United States. Homing in, we find that its source is the campus of the Massachusetts Institute of Technology on the banks of the Charles River near Boston. MIT was founded in 1861 by William Barton Rogers, a natural scientist who dreamed of establishing a new kind of independent educational institution relevant to an industrialised America. Rogers stressed the pragmatic and practicable and believed that professional competence was best fostered by coupling teaching with research and by focusing attention on real-world problems.

His legacy was one of the great intellectual powerhouses of the modern world, a university with the highest ambient IQ on earth. Although it was at first overshadowed by Harvard, its grander, older and more patrician neighbour, and looked down upon as a mere training school for engineers, nevertheless from the 1920s onwards MIT increasingly attracted the brightest and best of America's (and latterly the world's) scientists and engineers. In the middle decades of this century, the Institute became a seething cauldron of ideas about information, computing, communications and control. And when we dip into it seeking the origins of the Net, three names always come up. They are Vannevar Bush, Norbert Wiener and J. C. R. Licklider.

Vannevar Bush (1890–1974) was an electrical engineer and inventor of incredible ingenuity who spewed out patentable ideas which he rarely bothered to patent. He was one of those engineers who can turn their hands to virtually anything; and although he spent his professional life among the intellectual and political elites of the United States he never abandoned his homespun philosophical style which owed more to Will Rogers than to Santayana. His memoirs[5] evince a rough-hewn, innocent faith in the force of common sense and the American Way. Yet he made an indelible impression on those who studied or worked with him, and during the Second World War – after President Roosevelt had put him in overall charge of the country's scientific and technological effort – was one of the most powerful men in the United States.

Bush was born in Everett, Massachusetts and did his undergraduate and graduate studies at Tufts University, where his Master's thesis in 1913 included the invention of a device called the Profile Tracer, used in surveying work to measure distances over uneven ground. Profiles were usually obtained by using theodolites and required several men, careful measurement and a good deal of calculation afterwards before they could be drawn up. Bush set out to mechanise this process and constructed a weird contraption which looked like a vertical section through a low-slung perambulator. At either end was a bicycle wheel; in between was an instrument box containing gears, a rotating drum and a pen linked to a sensor. At the back was a handle which Bush used to push the device forward over the terrain to be profiled. As he progressed, a record of the elevation of the ground was traced on paper fixed to the revolving drum. It was, said Bush, 'quite a gadget', but it got him his Master's degree and a patent[6] though, alas, no fortune. As an intensely practical machine making intelligent use of mechanical devices, it perfectly exemplified his approach to problem-solving. And it was an intriguing forerunner of his later work on analog* computation.

In 1919, after a year with General Electric and a period teaching at Tufts, Bush joined the Electrical Engineering department at MIT and stayed there for twenty-five years, becoming Dean in 1932 and eventually Vice-President of the Institute. During that period he worked on a number of research projects connected in one way or another with information-processing. He developed, for example, a series of optical and photo-composition devices and a machine which rapidly selected items from banks of microfilm; but he was most celebrated in the 1930s for constructing a computer called the Differential Analyser.

The drive to construct the Analyser came originally from the needs of the electricity industry. In the years after the First World War, as long-distance power-lines brought electric light and power to more and more of the United States, there was an urgent need to understand how the evolving networks behaved – in particular, how they would be affected by surges, sudden line-failures and other events which could destabilise the system and cause it to fail. The difficulty was that the mathematical problems implicit in trying to analyse such systems were formidable. At the same time, given the huge investments involved, it was vital that engineers

---

*Because much of this story is based on U.S. sources I am using the American spelling rather than the British 'analogue'.

should be able to predict the operating characteristics and stability of the networks they were creating. But in order to do that they had to solve complex differential equations which were beyond the reach of purely mathematical analysis.

Some time in 1925, Bush and his students set to work on this problem. Drawing on some ideas first published by British scientists in 1876 they came up with a design for a machine which could function as what mathematicians call an 'integrator' and by 1928 had a working device capable of solving certain kinds of differential equation.

The Differential Analyser was essentially a gigantic *analog computer* – that is, a machine which represents variable physical quantities such as fluid pressure or mechanical motion by analogous mechanisms like variable-speed gears. It took up the entire length of a large laboratory and was inspired by the idea that many important dynamic phenomena in real life can be represented in terms of differential equations. In explaining this to audiences, Bush often used the analogy of an apple dropping from a tree. The thing we know about the apple is that its acceleration is approximately constant, and we can express this fact in mathematical symbols. In doing so we create a *differential equation* which, when solved, will give us the position of the apple at every point in time during its fall.

As it happens, this particular equation is very easy to solve. But suppose we want to include the effects of the resistance of the air in our differential equation? It's easy to do – just a matter of adding another term to the equation – but it has the effect of suddenly making the equation very hard to solve with pen and paper. Fortunately, though, it is easy to solve by machine. 'We simply connect together', Bush wrote, 'elements, electrical or mechanical gadgets, that represent the terms of the equation, and watch it perform.'[7]

Bush's Analysers (he built three versions in all during the 1930s) used replaceable shafts, gears, wheels and disks and required a great deal of setting up and maintenance. The computer had physically to be reconfigured for each equation. It thus did not 'compute' in the modern sense of the term, but rather *acted out* the mathematical equation it was seeking to solve. The final version of the Analyser was an enormous 100-ton electromechanical machine which could solve equations with up to eighteen variables and was as much sought after by the US military (for example, for ballistic calculations) as by scientists and engineers seeking answers to more innocent questions.

Bush left MIT in 1939 to become President of the Carnegie Institution. He was then recruited by Roosevelt to run the scientific side of what became the US war effort. When he arrived in Washington, Bush was mostly celebrated for his computing research. But although the Differential Analyser had stimulated a lot of thinking in MIT and elsewhere, it turned out to be, intellectually speaking, a dead end: it was the makers of digital, not analog, computers who would eventually inherit the earth.

The great irony is that the idea for which Bush will always be remembered sprang not from his computing work but from his interest in the problems of information management and retrieval. He had worked on these since the early 1930s and by 1939 had reached the point where he had sketched out the principles of a machine which would provide the mind with the kind of leverage that the steam engine had once provided for human muscle. Bush called it the 'Memex', and he was considering where to publish the first description of it when the war broke out. So he put the draft in his drawer, where it remained until 1945. Fortunately, he came back to it – and so will we. But that's for another chapter.

One of the people who worked with Bush on his Differential Analysers was another Tufts alumnus, Norbert Wiener (1894–1964), who had joined the MIT faculty in the same year (1919) as an instructor in mathematics. Wiener was someone for whom the label prodigy might have been invented. He was born in Columbia, Missouri of immigrant Jewish stock, and showed great promise from an early age. When he was eighteen months old, his nanny, amusing herself on a beach by making letters in the sand, noticed that he was watching her intently. As a joke she began teaching him the alphabet, and was astonished to find that within two days he had mastered it. He was a fluent reader by the age of three and by six had read some serious literature, including books by Darwin and other scientists.[8]

It's impossible to understand Norbert Wiener without knowing something of Leo, the remarkable, domineering father who shaped him. Leo was born in 1862 in Bialystok – then part of Russia, now in Poland – the son of a Jewish schoolteacher who sought to replace the Yiddish spoken in his community with literary German. After a sound classical education, Leo tried studying medicine and engineering but dropped out and eventually fastened on the idea of founding a vegetarian-socialist utopian community in the United States. These dreams, however, soon came unstuck: he arrived in the New World with twenty-five cents to his name and an urgent need to earn his living, which he did by working as a farmhand, a labourer, a janitor, a bakery delivery-man and a peddler.[9] From this humble start, however, Wiener Snr worked himself up to considerable eminence. He had a marvellous gift for languages and eventually became a schoolteacher in Kansas, married the daughter of a department-store owner in St Joseph, Missouri, moved to Boston and in 1896 was appointed an instructor in Slavic languages at Harvard. Fifteen years later, in 1911, he attained the heights of a Harvard chair – quite an achievement for a Jew in those days.

As a parent, Leo Wiener was a bully who rarely left anything to chance. He had strong ideas about how his children were to be educated. 'Instead of leaving them to their own devices,' he told a popular magazine once, 'they should be encouraged to use their minds to think for themselves, to come as close as they can to the intellectual level of their parents.' This required that parents should be constantly watchful of their words and actions. When in the presence of their children they should use only the best English, discuss only subjects of real importance and in a coherent, logical way; and make the children feel that they consider them capable of appreciating all that is said. (These stern specifications, however, did not include enlightening the child about his ethnic origins. Norbert was an adolescent before he discovered he was Jewish, and the shock of the discovery is movingly recounted in his autobiography.)

Much of Wiener's early schooling took place at Papa's hands. He was a tyrannical task-master. Every mistake had to be corrected as it was made. Lessons would start with father in a relaxed, conversational mode which lasted until Norbert made his first mistake. Then the loving parent was transformed into an enraged bully. By the end of many sessions the boy was weeping and terrified. 'My lessons often ended in a family scene,' he wrote in his autobiography. 'Father was raging, I was weeping and my mother did her best to defend me, although hers was a losing battle.'[10]

Norbert was understandably scarred by his relationship with his extraordinary father. Like many children of such a domineering parent, he nursed a chronic ambivalence towards him throughout his adult life. A dedication to Leo in one of his later books[11] is to 'my closest mentor and dearest antagonist'. And, writing in his autobiography of another child prodigy who cracked under the pressure, Norbert wrote: 'Let those who carve a human soul to their own measure be sure that they have a worthy image after which to carve it, and let them know that the power of molding an emerging intellect is a power of death as well as a power of life.'[12] Wiener bent under his father's pressure, but he did not break. In 1906, he enrolled at Tufts College, from which he graduated in 1909 with a degree in mathematics at the age of fourteen. He then went to Harvard as a graduate student and spent a year doing zoology before quitting because his physical clumsiness rendered him incapable of laboratory work. Undeterred, he changed to philosophy and completed his PhD at the age of eighteen.

Unsure of what to do, Wiener then went in the autumn of 1919 to Cambridge (England), where he studied under Bertrand Russell and G. H. Hardy. His relationship with the former was not entirely unproblematic. 'At the end of Sept.,' wrote Russell,

an infant prodigy named Wiener, Ph.D. (Harvard), aged 18, turned up with his father who teaches Slavonic languages there, having first come to America to found a vegetarian communist colony, and having abandoned that intention for farming, and farming for the teaching of various subjects . . . in various universities. The youth has been flattered, and thinks himself God Almighty – there is a perpetual contest between him and me as to which is to do the teaching.[13]

Wiener arrived in Cambridge around the time that Ludwig Wittgenstein set off for Norway, leaving Russell feeling bereft. And although the great man found Norbert personally repulsive, he was grateful, in Wittgenstein's absence, to have at least one precociously gifted student.[14]

From Cambridge, Wiener went to Göttingen in Germany to study with David Hilbert, one of the greatest mathematicians of the century. His first academic paper (in mathematics) was published just before the outbreak of the First World War. Discouraged from enlisting because of poor eyesight, Wiener drifted into a variety of occupations: teaching at the University of Maine (where he had difficulty maintaining class discipline), writing articles for the *Encyclopedia Americana*, enrolling as an engineering apprentice, working briefly as a journalist, and doing a stint as a mathematician calculating ballistic firing tables at the US Navy's Aberdeen Proving Ground in Maryland.

He joined MIT just as the Institute was beginning its great climb to scholarly eminence, and in the 1920s and 1930s did groundbreaking research in mathematics. Most academics would have been content to produce in a lifetime two or three papers of the quality that he rolled out by the score, but Wiener had a capacious intellect and an insatiable curiosity which led him constantly to stray into other fields. In the 1930s, for example, he joined a series of private seminars on scientific method organised by Arturo Rosenblueth of the Harvard Medical School. This was a remarkable group for its day: it deliberately mixed scholars from different academic disciplines, for example, and focused on 'large' subjects like communication processes in animals and machines. It also provided Wiener with a model for similar seminars which he would later organise himself. The participants were mostly young scientists at the Harvard Medical School and they would gather for dinner at a round table in Vanderbilt Hall. 'The conversation', Wiener recalled, 'was lively and unrestrained. It was not a place where it was either encouraged or made possible for anyone to stand on his dignity.'

After the meal, somebody – either a member of the group or an invited guest – would read a paper on some scientific topic, generally with a methodological slant. The speaker had to run the gauntlet of an acute, good-natured but unsparing criticism. It was a perfect catharsis for half-baked ideas, insufficient self-criticism, exaggerated self-confidence and pomposity. 'Those who could not stand the gaff', wrote Wiener, 'did not return, but among the former habitués of these meetings there is more than one of us who feels that they were an important and permanent contribution to our scientific unfolding.'[15]

After the Japanese attack on Pearl Harbour and the entry of the US into the war, Wiener began casting around for something useful to do to aid the war effort. Characteristically, his first idea was blazingly original. He sent Bush a design for a *digital* computer which incorporated all of the essential ingredients of the computers which eventually emerged towards the end of the war.[16] But Wiener was too far ahead of the game: nothing came of his idea and he cast around for something else to do. At the time, the two most important military projects for scientists and engineers were the atomic bomb project (based at Los Alamos) and research on ways of stopping German bombers which was beginning in a new laboratory – the Radiation Lab – that Bush had set up at MIT. This was the area Wiener chose.

His topic was control systems for radar-directed anti-aircraft guns. Effective anti-aircraft fire required a number of things: a good gun; a good projectile; and a fire-control system which enabled the gunner to know the target's position at all times, estimate its future position, apply corrections to the gun controls and set the fuse properly so that it would detonate the projectile at exactly the right instant. Wiener brought his formidable mathematical insight to bear on one aspect of this problem – that of using the available information about the location and motion of the aeroplane to make a statistical prediction of its future course. What he came up with was a mathematical theory for predicting the future as best one can on the basis of incomplete information about the past.[17]

Wiener's work on anti-aircraft fire brought him face to face with the central problems of feedback and control in so-called 'servomechanisms' – the machines used to control the guns. A servo-mechanism works by (1) comparing the actual orientation of the gun with its desired orientation (centred on the target), (2) feeding back a measure of this discrepancy (between where the gun is and where it ought to be) to an actuator (for example, a motor) which then (3) moves the gun in such a way as to eliminate the discrepancy. Wiener's work involved using mathematics to make an intelligent guess about where the target might be. The purpose of the servomechanism was to move the gun so that it was pointing at the predicted location when the target arrived there. It's simple in principle, but fiendishly difficult in practice. An anti-aircraft gun is a heavy object which builds up considerable momentum when it moves. This causes it to overshoot the desired orientation and necessitates another prediction and another gun movement – and so on and so forth. Automatic fire-control is thus an ongoing process.

But then Wiener noticed something which reminded him of the discussions he had had with Rosenbleuth and his neurophysiologists. In some circumstances, the gun-control system produced oscillations in which the weapon moved incessantly between two fixed points (a phenomenon known to engineers as 'hunting'). Wiener was suddenly struck by analogies between these pathological mechanical behaviours and the nervous tremors which Rosenbleuth and his colleagues had observed in human patients who had suffered neurological damage. And he had the genius to ask the question: were the underlying mechanisms the same?

Wiener discussed this ideas with Rosenbleuth, and both men came away convinced that they had stumbled upon a really big idea – that self-steering engineering devices using feedback loops for constant adjustment were analogous to human processes like picking up a pen or raising a glass of water to one's lips. The implications of this insight were stupendous, for it meant that the functioning of living organisms – including the human nervous system – might be amenable to mathematical analysis. For Wiener, the insight was particularly satisfying. His professional life had been spent as a mathematician with an insatiable curiosity about two other fields – engineering and biology. Now he had discovered that the two might be linked – and that his beloved mathematics formed the bridge.

It took him some time to work this out in a presentable way, which is why it was not until 1948 that he published his *magnum opus* – *Cybernetics: or Control and Communication in the Animal and the Machine*. The title came from the Greek word for 'steersman', and was intended to convey the idea of information-driven feedback control which Wiener perceived as being ubiquitous in both the mechanical and the animal worlds. It is a measure of his greatness that *Cybernetics* seems to us like a statement of the obvious. Reading it now, one is struck by its courtly, antique style, its apparently haphazard construction and the wildly fluctuating tone of the exposition. On one page Wiener writes with a scrupulous, literary clarity; but turn over and you are suddenly bombarded with mathematical symbols about unnormalised possibility densities and other arcana. It's the product of an intellect fizzing with ideas which have to be expressed somehow before their originator explodes in his exuberance. All in all, the thing's a mess.

And yet it was a sensational book in its day,[18] for it argued that there were unifying principles which spanned academic fields hitherto regarded as entirely separate. Henceforth anyone seeking to understand the world would have to deal in concepts like information, communication, feedback and control. And it argued that the greatest breakthroughs in human understanding were likely to come from the intellectual no-man's land between academic specialisms.

In a wonderful opening chapter, Wiener spells out his conviction that it is these 'boundary regions' of science which offer the richest pickings to the researcher. This is because they are ideas where the traditional methods simply don't work. When faced with a big intellectual puzzle, scientific disciplines tend to do two things – throw lots of man-hours at it; and divide up the task. This, for example, is the way genetics researchers tackled the task of mapping the human genome in the Sanger Centre near Cambridge. But, in subject areas which lie between specialist disciplines, this approach doesn't work. 'If the difficulty of a physiological problem is mathematical in essence,' Wiener reasoned, 'ten physiologists ignorant of mathematics will get precisely as far as one physiologist ignorant of mathematics, and no further.'[19] And if a physiologist who knows no mathematics works with a mathematician who knows no physiology, the first will be unable to state his problem in terms that the second can manipulate, while the second will be unable to put the answers in any form that the first can understand.

The implication was that a proper exploration of these blank spaces on the map of science would only come from truly interdisciplinary research done by teams of researchers – each a specialist in a particular field, but possessing a good acquaintance with neighbouring fields and accustomed to working together.[20]

Rosenblueth had returned to his native Mexico in the early 1940s, but as MIT got back to normality after the war, Wiener reactivated his idea of a regular multidisciplinary seminar. In the spring of 1948 he convened the first of the weekly meetings that were to continue for several years. Wiener believed that good food was an essential ingredient of good conversation, so the dinner meetings were held at Joyce Chen's original restaurant – now the site of an MIT dorm.

The first meeting reminded one of those present, Jerome Wiesner, of 'the tower of Babel', as engineers, psychologists, philosophers, acousticians, doctors, mathematicians, neurophysiologists, philosophers and others tried to have their say. But as time went on, says Wiesner, 'we came to understand each other's lingo and to understand, and even believe, in Wiener's view of the universal role of communications in the universe. For most of us, these dinners were a seminal experience which introduced us to both a world of new ideas and new friends, many of whom became collaborators in later years.'[21]

Some of Wiener's greatest intellectual contributions were concerned with the statistical interpretation of *time series* – sets of measurements of physical quantities taken at regular intervals over a period of time. He concentrated on developing mathematical techniques for extracting information from such data sets by 'smoothing', interpolating and predicting them. Given these interests, it was predictable that Wiener would be fascinated by the work of another MIT alumnus, Claude Shannon, who had done his PhD at the Institute but then moved to Bell Labs – the research arm of the telephone giant, AT&T.[22] For Shannon was obsessed with a similar problem – how to extract signals from noise in an electronic circuit like the telephone system.

You can see why AT&T would be interested in this kind of stuff. A long-distance telephone line is a communication channel through which electrical signals are propagated. But the line is subjected to interference – static, electromagnetic radiation and so forth – which gives rise to what engineers call *noise*. In audio terms, it's hiss. In visual terms it's the dancing dots on a television screen left switched on after the transmitting station to which it is tuned has ceased to broadcast.

The terrible thing about noise is that there's not much you can do to prevent it – it's a fact of electronic life. To the telephone engineer it's Public Enemy Number One. The longer the line, the weaker the signal gets, until eventually it is overwhelmed by the noise. The only way of dealing with this inexorable reduction in the signal-to-noise ration was to boost the signal at intervals by installing amplifiers, which was expensive in terms of both installation and maintenance costs.

Bell Labs attacked this problem in two ways. One was to invent a better, cheaper, more reliable amplifier – the transistor (which Bell researchers John Bardeen, Walter Brattain and William Shockley came up with in December 1947). The second approach was to try and get a mathematical handle on the process of communication. Shannon's great contribution was to develop the first rigorous model of the process, from which they inferred some general principles.[23] His model, in its original form, contained five elements, linked together in series – an *information source*, a *transmitter*, a transmission *channel*, a *receiver* and a *destination*. In time, the model was elaborated somewhat. The source was split into *source* and *message*, and components for encoding and decoding added.

The revised model looked like this: A *source* (a telephone user) uses an *encoder* (the telephone microphone) to transform a *message* (the words spoken) into electrical signals which are transmitted down a *channel* (the telephone network) until they reach a *decoder* (the earpiece of a telephone) which translates them back into sound waves capable of being understood by the *receiver* (the mind of the listener).

The Shannon model was important for several reasons. It changed communications theory from guesswork to science. It defined information in terms of reduction of uncertainty. It transformed the process of communication into something that could be empirically studied. It proposed quantitative measures of the effectiveness of communications systems. And it inspired the invention of the error-correction codes which make digital communications so much more robust and reliable than their analog predecessors – and which enable my modem to talk to the Net with such rapid-fire assurance.[24]

The Shannon model galvanised debate about communications not just because of what it explained, but because it also highlighted what it could not explain. How, for example, is it possible for humans to interpret messages which are apparently almost overwhelmed by noise? One explanation is that human communications rely very heavily on *redundancy*, that is to say superfluity of symbols. It has been claimed, for example, that most languages are roughly half redundant. If 50 per cent of the words of this book were taken away at random, you would still be able to understand what it was about (though it might look peculiar in places). So one way to ensure that the message gets through is to incorporate plenty of redundancy in it. This realisation was to prove very important in the design of communications systems.

But redundancy itself is not enough to explain the efficiency of human communication. One of Wiener's great insights was his realisation that *feedback* also plays an important role in the process. If we don't understand what someone says, we relay that fact back to them – by facial gesture, hand or head action, vocal interruption or whatever – and the speaker responds by slowing down, repeating or elaborating. Thus, even in the least mechanical of human social interactions, cybernetic principles are at work.

Wiener also understood very early the significance of computing machines. He had worked with Vannaver Bush on the early analog computers and (as we have seen) had himself sketched out a design for a digital machine in 1939. He also understood that the interesting question was not whether computers would 're-place' humans, but how people would interact with these 'think-ing machines' as they became more sophisticated. 'What functions should properly be assigned to these two agencies', he wrote towards the end of his life, 'is the crucial question for our times.'[25]

Wiener seems to have oscillated in his views of how computers would impact on people. Sometimes he viewed the machines as potential replacements for human brains; more frequently he seems to have visualised them as supplementing rather than replacing human skills in many areas of life. But he was alarmed that the only inference industrialists wanted to draw from the emergence of the digital computer was that it was the contempor-ary equivalent of the steam engine. 'There is no rate of pay', wrote Wiener, 'at which a United States pick-and-shovel laborer can live which is low enough to compete with the work of a steam shovel as an excavator.' The coming industrial revolution was similarly bound to devalue the human brain – at least in its simpler and more routine decisions.[26] In this Brave New World, the average human being of mediocre attainments or less would have 'nothing to sell that it is worth anyone's money to buy'. The only solution Wiener could envisage was 'to have a society based on human values other than buying or selling'.[27]

With his usual innocence, he tried to alert high officials in the US trade union movement to the emerging threat, but although he was received politely, his message fell on uncomprehending ears. The bosses of the AFL-CIO literally had no idea what this strange genius from MIT was talking about.

These were the ideas – about the ubiquity of feedback, the centrality of communications, the importance of control, the astonishing potential of computers, the need to call a truce in the turf wars between academic disciplines – that Wiener poured into the discussions over dinner at Joyce Chen's restaurant. They seem obvious, banal even, to us now, but in 1948 they were revolu-tionary. Putting them into the heads of young scientists and engineers was like lighting the fuses of a battery of rockets, some of which would fizzle out, some explode, and some soar to unimagined heights. Genius, Thomas Edison famously said, is 1 per cent inspiration and 99 per cent perspiration. With Norbert Wiener it was always the other way around.

He was one of those people who leave an indelible impression on all who encounter them. Physically untidy and clumsy, he was somewhat podgy, with a goatee beard. He habitually dressed in a three-piece suit and smoked cigars incessantly. He was usually extraordinarily generous with his time and ideas, especially to students, and yet occasionally behaved with the self-centred spitefulness of a thwarted five-year-old. He was entirely egalitarian in his behaviour, treating janitors and Nobel laureates alike. And he was famously absent-minded. Steve Heims relates a celebrated story about a student meeting him around midday and engaging him in conversation. Some time later Wiener asks, 'Do you remember the direction I was walking when we met? Was I going or coming from lunch?'[28] There are other accounts of him wandering into the wrong lecture hall and giving a lecture to a group of startled but enthralled strangers.

Wiener was very gregarious, with an insatiable need for conversation. His academic colleagues used to talk about the 'Wienerweg' – the route round the MIT campus he would take when he was working out a problem, was stuck or depressed or euphoric, or simply concerned about the political situation. One of his regular stops was at the room of Julius Stratton, a former student of his. Stratton eventually became President of MIT, but found that his new office also lay on the Wienerweg. 'It is customary', he recalled, 'when you come to the lobby of the President to ask if he was busy – not Norbert. He would just walk right in – whoever was there – he would interrupt and talk. This was not rudeness – he was just carried away.'[29]

One of the people who had his imagination fired by Wiener was a tall, sandy-haired Harvard scientist named J. C. R. Licklider. He was born in St Louis in 1915 and was a polymath in the Wiener mould, except that 'Lick' (as he was universally known) collected degrees like other people collect stamps. By the time he left college in 1937 he had undergraduate degrees in psychology, mathematics and physics. To these he rapidly added a Master's in psychology and a PhD in the psycho-physiology of the auditory system. He went to Harvard in 1942 as a research associate in the university's Psycho-Acoustic Lab and spent the later years of the war studying the effects of noise on radio reception. After the war, while back working at Harvard, he heard about Wiener's Tuesday evening seminars and signed up.

What fascinated Licklider was Wiener's emphasis on the human–computer relationship. At one point Wiener had conducted a small experiment to determine how the computer could aid him in his intellectual work. The study showed, he wrote, that 'almost all my time was spent on algorithmic things that were no fun, but they were all necessary for the few heuristic things that seemed to be important. I had this little picture in my mind of how we were going to get people and computers to really think together.'[30] By algorithmic, he meant tasks that were essentially procedural – complicated calculations, perhaps, or data sorting. Heuristic tasks, in contrast, were more open-ended, involved trial and error and the judicious application of rules which were not guaranteed to work. Implicit in this analysis was a picture of computers as the tools which would do the algorithmic stuff, thereby improving his ability to do the (heuristic) work requiring the exercise of human judgement.

This co-operative aspect of the human–computer relationship was what struck Licklider most forcefully. The trouble was that it was difficult at the time to have a relationship of any kind with a computer. In the early 1950s they were vast machines with which one interacted by submitting 'jobs' (programs to be run) encoded on decks of punched cards. The programs were run in batches, and all the emerging computer manufacturers regarded batch-processing as an entirely sensible and efficient way to use scarce computational resources.

Interacting with a batch-processing machine could be a pretty dispiriting affair. It was my first experience of computing in the mid-1960s and it damn nearly put me off it for life. The machine we used was an IBM 1401 which sat in an air-conditioned suite, attended by operators and clerks. No student was ever allowed into the hallowed 'machine room'. We typed our programs line by laborious line on a machine with a QWERTY keyboard. With every key press came a 'thunk' as the machine punched a hole in a card. At the end of each line, it spat out a completed card. A deck of punched cards constituted a program.

Having written your program, you filled out a paper from with your user-id, course code and other details, placed form and deck in a transparent plastic bag and deposited the lot in a tray with other decks waiting to be run. At periodic intervals, an operator would appear, pick up the decks and spirit them away to the machine room. Hours later, the operator (or his counterpart from a later shift) would reappear with your deck and some computer printout giving the results of the program run or (much more likely) an error report resulting from a bug in your code. Then it was back to the thunk machine to reassemble the deck, submit, wait . . . and so on and so on (debugging is an iterative process, remember) – until you were seized with an overwhelming desire to strangle the entire board of directors of IBM with your bare hands.

Something similar faced Licklider at Harvard, so it was not surprising that in 1950 he was lured down the river to the superior computing resources available at MIT. He first joined the Institute's Acoustic Laboratory, but the following year, when the Cold War really began to chill and MIT set up the Lincoln Laboratory as a facility specialising in air defence research, he moved over to head up the lab's human engineering group – a transition he described as becoming the 'token psychologist' in a group of scientists and engineers, with access to the most advanced electronics in existence at the time.

The wondrous kit in question was provided initially by the Whirlwind project. This was a research programme directed by Jay Forrester which had begun in 1947 and led to the design of computers with which one could directly interact. It had its origins in a proposal to build an analog aircraft simulator for the Air Force, but the realisation of how cumbersome the requisite technology would be led to a redefinition of the project as one requiring digital computers. In 1949, as the Cold War intensified, Whirlwind was given new impetus, a new sponsor (the US Navy) and virtually limitless funding.[31]

Then the Air Force decided that it needed a computerised electronic defence system against the threat of Russian bombers armed with nuclear weapons. Whirlwind was adopted as the prototype and test bed for the new system, which was christened SAGE (for Semi-Automatic Ground Environment) and designed to co-ordinate radar stations and direct fighters to intercept incoming planes. SAGE consisted of twenty-three 'direction centers', each with a computer which could track as many as 400 planes and distinguish enemy aircraft from friendly ones.

Whirlwind and SAGE were pathbreaking projects in almost every way. Whirlwind led to the invention of several key technologies like 'core memory' – the first type of Random Access Memory (RAM) which did not require vacuum tubes. And it gave people their first impressions of truly interactive computing. 'It took 2500 square feet,' recalled Ken Olsen, the founder of Digital Equipment Corporation, who worked on the project as a researcher. 'The console was a walk-in room as big as this loft here. But in a real sense it was a personal computer and did personal computer things.'[32]

Whirlwind was a machine one walked *into*. What now fits comfortably into my laptop, in those days required a whole room for the control equipment alone. Programming was a primitive business carried out with the aid of ruled coding forms on which one wrote out numerical codes. But when Whirlwind's early users entered the belly of the beast armed with their code, a unique experience awaited them. For their assigned block of time, typically fifteen minutes, *the machine was theirs and theirs alone!* 'What makes the Whirlwind computer important in the history of the personal workstation', explains one of its users, 'is the fact that it was the first really high-speed machine with powerful interactive display capabilities.'[33]

SAGE, for its part, was even more gargantuan. The computer itself used 55,000 vacuum tubes and weighed 250 tons. The system had to handle many different tasks at the same time, and share central processor time among them. It gathered information over telephone lines from a hundred radar and observation stations, processed it and displayed it on fifty screens. The 'direction centers' were also linked to each other by telephone lines.

SAGE was an enormous project which required six years of development, 7,000 man-years of programming effort and $61 billion in funding. The phone bill alone ran to millions of dollars a month. (National Security never came cheap.) But the most important thing about SAGE was the word 'semi-automatic' in its title: it meant that human operators remained an integral part of the system. They communicated with the computer through displays, keyboards and light-guns. They could request information and receive answers in seconds. SAGE might have been a monster, but it was also 'one of the first fully operational, real-time interactive computer systems'.[34] To Licklider, it was a breathtaking illustration of what humans and computers could do together.

As he got deeper into the work of the Lincoln Lab, Licklider also got drawn into programming. He was introduced to this addictive sport by Wesley Clark, who was then working on the TX-2 computer – one of the first built using transistors and a precursor of the PDP[35] line of computers later manufactured by Digital Equipment Corporation. 'I remember when I first met Lick', said Clark,

> he was in the same basement wing of the Lincoln Laboratory in which the TX-2 was operating. And I wandered down the hall one time, and off to the side of this dark room, way near the end of the hall, sort of . . . I think that's where the stock room was . . . or part of it anyway. And off to the side was this very dark laboratory and I went in, and after probing around in the dark for a while I found this man sitting in front of some displays and doing some things. And he was doing some kind of . . . a piece of psychometrics, or preparing an experiment for such a study, perhaps by himself. And we began to chat, and he was clearly an interesting fellow. And I told him about my room, which was sometimes dark down at the other end of the hall, and invited him to come down and learn about programming, and he subsequently did that.[36]

The thing that fascinated Licklider about the TX-2 was its

graphics display. Most computers of the time communicated with the user (and vice versa) by means of a typewriter-like device called a teletype. The TX-2, in contrast, displayed information on a screen. Hypnotised by the possibilities of this, Licklider became hooked on computing and took to spending hours at a time with the machine. He had seen the future – and it worked for him.

What Licklider gained from these subterranean sessions was a profound conviction about the importance of *interactive* computing. He eventually articulated this philosophy in 'Man–Computer Symbiosis' – one of the seminal papers in the history of computing, which was published in an engineering journal in March 1960.[37] Its central idea was that a close coupling between humans and computers would result in better decision-making. In this novel partnership, computers would do what they excelled at – calculations, routine operations, and the rest – thereby freeing humans to do what they in turn did best. The human–computer system would thus be greater than the sum of its parts.

The idea for the paper did not come from any particular piece of research. It was, said Licklider, just a statement about the general notion of analysing work into creative bursts and routine programmable tasks that you could get to get a computer to do. This insight was partly based on experiments he had conducted on himself. 'I tried to keep schedules [to] see how much time I spent doing what,' he recalled, 'and I was pretty much impressed with the notion that almost all the time I thought I was thinking and working, I was really just getting in the position to do something.'[38]

Lick had also been very struck by an experiment he had carried out with one of his graduate students, an electrical engineer named Jerry Elkind. Elkind had done some experiments which yielded copious amounts of data and he was convinced there were some relationships hidden in the numbers. So he plotted the data on sheets of paper – and wound up with a large sheaf of pages, with the result that he still couldn't see what was going on. 'So', recalled Licklider,

we put big heavy blobs wherever there was a datum point, and went down to the Sloan building where I happened to have an office at the

end of a little mezzanine where you could stand and look down on
the floor below. So we redirected traffic a little bit, and put all these
graphs down there, so we had a hundred or so sheets of graph paper.
Then it was obvious what was going on. And I was pretty much
impressed. That happened frequently: you do a lot of work, you get
in a position to see some relationship or make some decision. And
then it was obvious. In fact, even before you could quite get finished,
you knew how it was going to come out. 'Man–Computer Symbiosis'
was largely about ideas for how to get a computer and a person
thinking together, sharing, dividing the load.[39]

Licklider's conception of *symbiosis* – 'co-operative living together
in intimate association, or even close union, of two dissimilar
organisms' – in the context of the relationship between humans
and machines gives some idea of how radical he was. Remember
that he was writing at a time when most people thought that
computers would simply be superior kinds of 'number-crunchers' –
that is, calculators. The metaphor came from biology. 'The fig tree',
he wrote,

> is pollinated only by the insect *Blastophaga grossorum*. The larva on
> the insect lives in the ovary of the fig tree, and there it gets its food.
> The tree and the insect are thus heavily interdependent: the tree
> cannot reproduce without the insect; the insect cannot live without
> the tree; together they constitute not only a viable but a productive
> and thriving partnership.

The amazing thing about Lick's idea is that it was a vision not of
the kind of relationship which exists between, say, robots working
on an assembly line and their human overseers, but of something
much more intimate. His hope was that 'in not too many years,
human brains and computing machines will be coupled together
very tightly, and that the resulting partnership will think as no
human brain has ever thought and process data in a way not
approached by the information-handling machines we know
today'.[40] In a lecture delivered around the same time, Licklider
predicted that 'In due course [the computer] will be part of the
formulation of problems; part of real-time thinking, problem-

solving, doing of research, conducting of experiments, getting into the literature and finding references . . .'[41] And, he added, in one of the great throwaway lines of computing history, it 'will mediate communication among human beings'.

The difficulty was that 'man–computer symbiosis' required more flexible computers than the batch-processing monsters current at the time. Bringing computers as partners into the thinking process means that interactions with them had to happen in 'real time' – something that batch-processing specifically precluded, because it required that data-processing went on at a pace convenient to the machine (and its keepers) rather than to the user. Here the fact that Licklider was based at MIT again proved crucial, because he was not alone in seeking to escape from the tyranny of the batch queue. Others, too, sought liberation. Their Holy Grail was something called 'time-sharing'.

Time-sharing in the modern sense[42] goes back to 1957 when John McCarthy, a young computer scientist from Dartmouth College, arrived at the MIT Computation Center on a Sloan Foundation fellowship. McCarthy is nowadays celebrated as one of the founding fathers of Artificial Intelligence, but in 1957 he was obsessed with what seemed to him an obvious idea – that of modifying the operation of an expensive computer so that it could serve many users at the same time. He wanted 'an operating system that permits each user of a computer to behave as though he was in sole control of the computer, not necessarily identical with the machine on which the program is running'.[43] The method of achieving this would be for the machine to cycle rapidly between each user, thereby giving each one the illusion of being the only person using it.

McCarthy's initial stab at the problem involved installing a relay (an electro-mechanical switch) which would enable a keyboard terminal to interrupt the processing of an IBM 704 mainframe computer whenever anyone typed a character on the terminal. This was crude, but it highlighted the central problems involved – how to devise a non-disruptive method of interrupting the processor as it cycled between tasks; and how to devise efficient scheduling methods for allocating resources between users.

In 1960, the MIT Administration appointed a committee to make recommendations on the Institute's long-term computing strategy, and McCarthy was one of the academics invited to join, though by this stage he was on his way back to his first love, Artificial Intelligence. In the spring of the following year he was invited to lecture to a symposium celebrating MIT's centenary and used this unexpected pulpit to explain why time-sharing was vital to the future of computing.[44]

Although there seems to have been some internal disagreement about the best way to proceed, the record shows that time-sharing rapidly became a major research and development obsession at MIT. By November 1961, Fernando Corbato and several colleagues had a four-terminal system working off an IBM 709 mainframe.[45] By 1962, when the Computation Center received its new IBM 7090 machine the Compatible Time Sharing System (CTSS) project was in full swing. Time-sharing was an idea whose time had come – which is why it spread with such rapidity through the research computing world. And why it persuaded those of us who had been repelled by the iniquities of batch-processing to give computing a second chance.

The most intriguing thing about time-sharing as a technology was its social impact. When I went to Cambridge in 1968, the main computer in my laboratory was an English Electric batch-processing machine. We coded our programs (in a language called ALGOL) by punching holes in paper tape and then feeding the resulting reels through a tape-reader. The computer sat in a large air-conditioned room and had a full-time operator who could be capricious in the way all absolute rulers are. If your programs took a lot of processor time – and you paid due obeisance to the machine's keeper – you could get permission to run it through the night. Among the small group of students who used this antique facility there was the kind of camaraderie you find when people are confronted with a common adversary, the fellowship of the bus queue. But it was no more than that.

There was, however, another group of students in the lab who also did a lot of computing, but not on our machine. Instead, they used the university's time-shared mainframe, an ancient behemoth appropriately called Titan which was housed in the Computing Laboratory on the Old Cavendish site. Initially, they communicated with this device using teletype machines which clattered like telex machines in 1950s movies and later via consoles which had screens and keyboards. Gradually it dawned on me that these students were different in some way from us. They were cliquey and gossiped more. When in our presence at the laboratory ritual of afternoon tea they talked like people who were in on some secret joke. They exchanged electronic messages in some mysterious way. They were more co-operative towards one another. And they were in love with the machine which kidded them into thinking it was all theirs. Some of them, indeed, were potty about it: when eventually the university decommissioned Titan in 1973 there was a wake attended by its mournful users, many of whom took away pieces of the hardware as personal keepsakes.

To a young engineering student like me, this seemed strange. To an anthropologist, however, it would have been entirely predictable. For what I was observing was the way in which time-shared machines created a sense of *community* among their users. Not surprisingly – given their head-start – the folks at MIT had long ago twigged this. Before time-sharing, computer users were often isolated individuals who didn't even recognise one another except as members of the same queue. But as soon as the technology became available its users rapidly got to know one another, to share information, to seek and give help. 'The time-sharing computer system', wrote two MIT researchers at the time,

> can unite a group of investigators in a cooperative search for the solution to a common problem, or it can serve as a community pool of knowledge and skill on which anyone can draw according to his needs. Projecting the concept on a large scale, one can conceive of such a facility as an extraordinar[il]y powerful library serving an entire community – in short, an intellectual public utility.[46]

None of this was lost on Licklider, of course. Nothing was ever lost on Licklider. He soaked up knowledge and information like a sponge, and transmuted them into ideas which energised the thoughts of others. His years at MIT transformed him utterly, from a prodigiously clever experimental psychologist to someone with a compelling vision of how computing technology could improve the human condition. From Wiener he got the idea of the centrality of the human–computer relationship; from Bush he got a sense of the importance of finding ways of managing the information explosion which threatened to overwhelm humanity; from Whirlwind, SAGE and the TX-2 he got his first inklings of what 'man–computer symbiosis' might be like; and the time-sharing experiments prodded him into realising that computing could not only empower individual decision-makers but also engender new kinds of intellectual communities. By the early 1960s, he was writing memos to his friends in what he called the 'Inter-Galactic Computer Network', by which he meant a vast, time-shared community.

Licklider left MIT in 1962 with the ideas needed to create the future. All he needed to make it happen was some institutional leverage. And, as luck would have it, the Russians had already laid it on.

5:
Imps

> The process of technological development is like building a cathedral. Over the course of several hundred years new people come along and each lays down a block on top of the old foundations, each saying 'I built a cathedral'. Next month another block is placed atop the previous one. Then along comes an historian who asks, 'Well, who built the cathedral?' Peter added some stones here, and Paul added a few more. If you are not careful you can con yourself into believing that you did the most important part. But the reality is that each contribution has to follow onto previous work. Everything is tied to everything else.
>
> Paul Baran[1]

Like all great systems-engineering constructs, the Internet did not originate in one blinding, 'Eureka!' moment. But if one had to put a finger on the spark that lit the fuse, one would have to say it happened on 4 October 1957 – the day the Soviet Union launched into orbit a bleeping football called Sputnik.

To say that Sputnik gave rise to some concern in the United States would be the understatement of the century. The truth is that the US went apeshit. The launch rattled even Dwight D. ('Ike') Eisenhower, the thirty-fourth President of the United States, normally a calm and phlegmatic man. Sputnik was even more painful for Ike because he and his advisers had long agonised about how to make federal funding of advanced research and development (R&D) more effective, and in particular how to elevate it above the vicious turf wars which raged between the three branches of the armed services.

It's a truism of military life that victory is usually assured the moment the various branches of the armed services begin to hate the enemy more than they detest each other. In the years after the Second World War, each arm of the US military had spent countless billions of dollars on advanced research into missiles and weaponry. And most of these programmes were duplicated. If the Air Force had nuclear weapons, then the Navy had to have its own missile programme. And if the Navy and the Air Force had rockets, then the Army had to have some too. And yet, despite all this expenditure on lethal rocketry, the country had not been able to launch so much as a tennis-ball into space. Sputnik provided vivid proof of the extent to which America had been humiliated by the Soviet Union, a country that – with the sole exception of aerospace products – apparently couldn't manufacture a decent garden spade.

In his 1958 State of the Union Message, Eisenhower went straight to the heart of the problem. 'I am not here today to pass judgement on harmful service rivalries,' he said. 'But one thing is sure. America wants them stopped.'

The Advanced Research Projects Agency (ARPA) was Ike's answer to the turf wars – a federal agency which would have overall control of the most advanced military R&D projects. It was set up within the Pentagon (the headquarters of the US Department of Defense) early in 1958 with a vice-president of General Electric as its first Director and half a billion dollars in its kitty, which at that time was serious money. Initially ARPA had responsibility for all US space programmes and advanced strategic missile research, but was nearly still-born because later the same year the National Aeronautics and Space Administration (NASA) was established and given all of ARPA's aerospace brief, not to mention most of its budget.[2]

It looked like an everyday story of life in the Washington bureaucratic jungle. Government gets Big Fright, declares existing agencies inadequate, sets up new agency; but old agencies regroup, play bureaucratic hard-ball and eventually crush plucky little new-comer. Yet it did not turn out like that, mainly because the outgoing ARPA Director asked his staff to draw up a paper redefining the Agency's mission. They re-engineered ARPA as an outfit which would take on 'blue skies' research and tap into the enormous pool of talent in American universities crying out for some national body to invest in 'high-risk, high-gain' research.

The renaissance of ARPA began under its second Director, an Army general called Austin Betts, and continued under his successor, Jack Ruina, the first scientist to head the Agency. Ruina was an astute manager with a laid-back style who steadily increased ARPA's budget, largely by cultivating the powers-that-be in the US defence establishment, and by funding research on subjects which were of most interest to them.

Initially this involved anti-missile defence and technology for detecting nuclear tests, but gradually the brief widened. The military, for example, began to realise the importance of behavioural science and asked Ruina to mount a research programme in this area. While he was still ruminating on how to do this, the Air Force started to look for an institutional home for a time-sharing computer project based at System Development Corporation (SDC) in California which had become surplus to requirements, and their gaze alighted on Jack Ruina's little empire.

Seeking to kill two birds with a single stone, Ruina went looking for someone who was interested in both behavioural science and time-shared computers. In the autumn of 1962 he came up with the names of two people – Fred Frick of Harvard, and J. C. R. Licklider of MIT – and invited both to come to Washington for a talk. Would either of them be willing to head up a behavioural science research programme? he asked. 'We were both interested in what he was talking about,' recalled Licklider, 'but neither one of us wanted to leave what we were doing.' Ruina then sent the reluctant pair to see Gene Fubini, an aide to the Assistant Secretary of Defense. Licklider was impressed by Fubini: 'He was an immigrant from Italy – a wonderful European technical education; very sharp, incisive, impatient kind of guy; quite eloquent, and really dedicated to this job at the Pentagon. He made Frick and me feel that the world needed at least one of us and we should do it.'[3]

In reality, Fubini was pushing at an open door. All he needed to

do was to persuade Lick to think out loud. 'When we got talking,'
recalled Licklider,

> I started to wax eloquent on my view . . . that the problems of
> command and control were essentially problems of man computer
> interaction. I thought it was just ridiculous to be having command
> control systems based on batch processing. Who can direct a battle
> when he's got to write the program in the middle of the battles?[4]

Fubini agreed wholeheartedly with this and so did Ruina. Licklider
went on to argue the need for time-shared, interactive computing.
It would undoubtedly have applications in command and control
systems – and therefore appeal to the military.

In the end, Licklider and Frick tossed a coin to decide who should
go to Washington, and Licklider won. Ruina offered him a job
heading up ARPA's Command and Control Division, with the
added incentive that he could also set up a new behavioural
sciences division. Lick arrived at the Agency on 1 October 1962
and hit the ground running. He had two overriding convictions.
The first was that time-sharing was the key computing technology;
the second was that the best way to make progress in research was
to find the smartest computer scientists in the country and fund
them to do whatever they wanted.

These were radical ideas. (The second one still is.) Up to then
ARPA had funded computing research mainly via the R&D departments of computer companies. Licklider's MIT experience had told
him that this strategy would lead nowhere. After interviewing
executives in one of the biggest of the companies, he observed
that its operations were entirely based on batch-processing. 'While I
was interested in a new way of doing things,' he said, 'they were
studying how to make improvements in the way things were done
already.'[5]

There was an ambiguity about Licklider's terms of reference
which he exploited ruthlessly. Ruina thought he'd appointed him
to establish behavioural science as a research priority and to take
responsibility for the SDC time-shared project. The Department of
Defense was under the impression that he'd been hired to run
Command and Control. Licklider creatively redefined all his

supposed responsibilities so that they mapped on to his obsessions.

> Every time I had the chance to talk, I said the mission is interactive
> computing. I did realize that the guys in the Secretary's office started
> off thinking that I was running the Command and Control Office,
> but every time I possibly could I got them to say interactive
> computing. I think eventually that was what they thought I was
> doing.[6]

Licklider's encounters with the batch-proceeding fanatics in the
mainframe computer industry made him determined that ARPA
funding for advanced research in the field would go to places which
understood the interactive message – which effectively meant
universities and a handful of labs linked to them. Things rapidly
got to the point where, if you wanted to buy a computer on an
ARPA project, you had to make sure it was a time-shared one –
which is how the Agency came to fund six of the first twelve
general-purpose time-sharing systems ever developed.

It was estimated that, in the years that followed Licklider's stint at
the Agency, 70 per cent of all the funding for computer science
research in the United States came from ARPA, and much of it
followed the path set by him in 1962. It was on his initiative,[7] for
example, that the Agency funded Douglas Engelbart's research
programme at the Stanford Research Institute which led to the
invention of the 'mouse' as a pointing device – and thus to the first
glimmerings of the Apple Macintosh and Microsoft Windows user
interfaces. Lick's determination to fund excellent people wherever
they were to be found led him to provide massive support to West
Coast labs at Berkeley and Stanford whose students, steeped in
visions of interactive computers, contributed to the development of
Silicon Valley and that ultimate manifestation of the intimate
machine – the personal computer. 'But the crown jewel of Licklider's crusade', wrote one observer, 'was his initiation of the events
leading to the Internet. Had Licklider not altered course, odds are
that there would be no Internet and the seemingly silly musings of
Wiener would, in hindsight, appear prescient.'[8] Though Licklider
left ARPA in 1964 to return to MIT after only two years in post, he

had an incalculable impact on the Agency. He signalled the demilitarisation of his department by changing its name from Command and Control Research to Information Processing Techniques Office (IPTO). His brief period in office set the scene for virtually everything that followed – the emphasis on interactive computing, enabled by time-shared systems; the utopianism which maintained that computer technology held out the promise of a better world; the belief that important synergies could be gained by devising ways in which people and machines could work together; and the conviction that the only way to get magic from research funding is to give bright people their heads.

The next significant development in the story came in 1965 when Bob Taylor, a Texan who in due course was to mastermind not one but two revolutions in computing, was appointed ITPO Director.[9] Taylor had begun his academic career in the same field as Licklider – psycho-acoustics – and then moved steadily into computing. After a few years as a systems engineer in the aerospace industry, he joined the Office of Advanced Research and Technology at NASA headquarters, where he worked from 1962 to 1965.

In late 1962, Licklider, then ensconced at ARPA, had set up an unofficial committee of people in federal agencies who were funding computer science research, and Taylor (who had been handing out NASA research grants to computer scientists like Douglas Engelbart) was invited to join the committee. Early in 1965, Licklider's successor as IPTO Director, Ivan Sutherland, asked Taylor to leave NASA and join ARPA as his deputy. 'The reason I moved from NASA', explained Taylor,

> [was] fundamentally because over time I became heartily subscribed to the Licklider vision of interactive computing. The 1960 man–computer symbiosis paper that Licklider wrote had a large impact on me . . . The Vannevar Bush work 'As We May Think' published in *Atlantic Monthly* 1945 had an influence . . . Some of the work of Norbert Wiener in the late 1940s also had some influence. I was interested in the junction of these three different themes. My graduate research had been concerned with how the nervous system

works. Over time I began to feel that the kind of computing research that Licklider was suggesting was going to be much more reward-ing . . . So I gradually got more and more interested in computer research and less and less interested in brain research.[10]

Within a few months, Sutherland himself had gone, and Taylor suddenly found himself catapulted into one of the most powerful jobs in the research firmament.

ARPA was one of the oddest bodies ever to fund advanced research. For one thing, it was small, efficient, decisive and incredibly unbureaucratic. 'I have a theory', mused Ivan Sutherland once, 'that agencies develop arthritis as they grow older.' But ARPA was young – and staffed by a different kind of animal from that normally found in Washington.

> The directors traditionally had been found outside DoD [Department of Defense]. They were not bureaucrats, but were typically university folk who were coming to try and do some reasonable technical job for the government with the minimum of fuss. There really was not very much fuss in ARPA. In fact, to get something done, what I had to do was convince myself and convince my boss. He apparently had spending authority.[11]

ARPA was also marked out by the fact that it had at its disposal massive amounts of money. And although it was physically and organisationally based in the Pentagon – the heart of the military-industrial complex – it was not oppressively burdened by military priorities. 'We were certainly a Defense Department agency,' says Taylor, 'but I never received any guidelines that the research that I sponsored should have a specific military connection.' What he was continually trying to do was to select research problems which, if solved, would have massive spin-off benefits.

> We were not constrained to fund something only because of its military relevance. Without being told by anyone, we would make connections between the research problems that we were attacking and the technology limitations that were faced by the Defense Department. But those problems that were faced by the Defense

Department were also faced by a wide variety of other endeavors of
the United States and, in fact, the world.[12]

One of the first things that struck Taylor when he took over as
IPTO Director was that his room in the Pentagon contained
teletype terminals to each of the three ARPA-supported time-
sharing systems – at MIT, SDC at Santa Monica (the machine that
had prompted Jack Ruina to recruit Licklider) and the University of
California at Berkeley.

> Three different terminals. I had them because I could go up to any
> one of them and log in and then be connected to the community of
> people in each one of these three places . . . Once you saw that there
> were these three different terminals to these three distinct places the
> obvious question that would come to anyone's mind [was]: why
> don't we just have a network such that we have one terminal and we
> can go anywhere we want?[13]

To Taylor the fact that the three machines couldn't communicate
represented a huge waste of federal resources. IPTO-funded researchers were demanding more and more computing power.
Remember that computers were then very expensive items –
typically costing anything from $500,000 to several million dollars.
Yet they were all incompatible – they could not even exchange
information with one another, and there was no way in which
researchers at one ARPA-funded site could access the (ARPA-
funded) computing resources at another without physically relocat-
ing. At the same time, a federal agency like ARPA could not prevent
researchers from buying computers from different manufacturers
because all US firms had to be given equal opportunities to tender.
The obvious thing to do was to link all these incompatible
machines, somehow. 'So', recalled Taylor, 'in February of '66 I
went into the Director of ARPA's – Charlie Herzfeld – office and told
him about this idea. And he liked it right away and after about
twenty minutes he took a million dollars out of someone else's
budget (I don't know whose to this day) and put it in my budget
and said, "Great. Get started!" '[14]

On such hinges does history turn: from zero to a million bucks in twenty minutes. Years later the myth spread that what drove ARPA to build the world's first computer network was the desire to provide a communications system that could survive a nuclear attack on the United States. The record suggests otherwise: Bob Taylor simply wanted to make the taxpayer's dollar go further.

It was one thing to get the money for a network, but quite another to get a guy who could make it happen. But, as luck would have it, Taylor had someone in mind. His name was Lawrence G. Roberts, and he worked at – where else? – MIT.

Larry Roberts was one of those kids who goes to university and never seems to leave. Finding himself under-stimulated by his undergraduate studies at MIT, he got a vacation job at the Institute's Computer Center helping to convert an automatic tape unit attached to the Center's IBM 704 machine. It was just a summer job but Roberts's first encounter with computers. He wrote programs in hexadecimal codes[15] and remembers the experience as 'painful'.[16]

Nevertheless, he was hooked. The following year, the TX-2 computer which had been built at the Lincoln Laboratory (and on which Licklider had learned to program) was transferred to the Institute proper. The TX-2 was the first all-transistor computer ever built. It filled a room, but was still small for those days. And it was the first machine of its kind ever to be available for online use by individuals. 'They did not have any plan for how their people were to use it,' said Roberts, 'so I, being around there, started using it . . . I was around the area . . . When I would go to the computer area, I started working with it. The first year that it was there I spent 760, or something, hours on-line working on the machine, which is, for that era, almost unheard of.'[17]

In fact, Roberts wrote most of the operating system – the internal 'housekeeping' program which managed the computer – for the TX-2, as well as some compilers for various programming languages. In addition he did a Master's on data compression and a PhD on the perception of three-dimensional solids. And he even found time to write a handwriting-recognition program based on neural networks which was way ahead of its time. When he embarked on his doctoral research, Roberts had become, *de facto*,

the world's leading expert on the TX-2. And when Wesley Clark
and Bill Fabian – the two guys responsible for the machine –
abruptly left the Lincoln Lab after a dispute about whether they
were allowed to bring a cat on to the premises, Roberts found
himself in charge of the project even though he was officially still a
graduate student.

Roberts says that he first began to think about networking issues
around 1964, when – along with others from the MIT time-sharing
community – he was invited to a conference on the future of
computing held in Homestead, Virginia and hosted by the Air
Force. Many of the conversations between the attendees (who
included Licklider) centred on the chronic incompatibilities which
then existed between different computer systems. 'We had all of
these people doing different things everywhere,' Roberts recalled,
'and they were all not sharing their research very well.' Software
developed for one machine would not run on another.

> So, what I concluded was that we had to do something about
> communications, and that really, the idea of the galactic network
> that Lick talked about, probably more than anybody, was something
> that we had to start seriously thinking about. So in a way networking
> grew out of Lick's talking about that, although Lick himself could not
> make anything happen because it was too early when he talked about
> it. But he did convince me it was important.[18]

In interviews and in his version of ARPANET history,[19] Roberts
implies that the Virginia meeting was something of an intellectual
watershed for him. This may well be true, but external evidence
that his interests shifted radically as a result of the conference is
harder to come by. He returned to MIT and continued to conduct
the (ARPA-funded) graphics research which at that time seemed to
be his main passion.

In 1965, however, Bob Taylor asked him to take a brief diversion
from computer graphics and work on 'a specific, one-time, short
experiment' to test Taylor's concerns about the noise and speed of
telephone lines carrying bits over a substantial distance. 'I wanted a
cross-country test between the TX-2 and another machine,' recalls
Taylor, 'so I arranged with System Development Corporation in

Santa Monica California to be the West Coast node (my office was supporting both their Q-32 work and the TX-2 work, so the arrangement was easy).'[20]

There is some disagreement about the genesis of this experiment. Hafner and Lyon maintain that the idea came from Thomas Marill, a psychologist and computer scientist who had worked with Licklider.[21] Bob Taylor disagrees, 'There was no interaction between me and Tom,' he says. 'Tom has claimed he sent a proposal to ARPA to build a network. I doubt it. I never saw it nor heard of it.'[22]

Taylor's intention was that Larry Roberts would run the test, but in the end the job was subcontracted by the Lincoln Lab to Marill's company, Computer Corporation of America. Whatever its provenance, the experiment was ambitious for its time – it involved linking two very different computers at opposite ends of the continent via a Western Union-leased telephone line. In the experiment, the two machines did not exchange files, but they did pass messages. The results were, to say the least, mixed. Taylor's fears were confirmed: the connection worked, but the reliability and response of the link were disappointing.[23] Networking on a continental scale was not going to be easy.

Building Bob Taylor's network would push ARPA into new territory, because up to then the Agency had handed out money and other people had done the work. In constructing the proposed network, however, it would be *driving* the project, rather than allocating funding to laboratories for particular kinds of advanced research and letting them get on with it. Conscious that he needed new skills and resources in his office, Taylor started looking for someone to manage the network project and he fastened quickly upon Larry Roberts. In February 1966 he invited him to become the programme manager for the project. Roberts turned him down flat. He did not, he said, wish to become a Washington bureaucrat.

Taylor, however, is not the kind of guy who gives up easily. 'I tried all during '66 to try to get him to change his mind,' he recalled, 'and I continually failed.'

Then in October or November of '66 I had an idea and I went in to see Charlie Herzfeld, the Director of ARPA. And I said, 'Charlie, is it true that ARPA supports 51 per cent of Lincoln Lab?' I said, 'Would you call the Director of Lincoln Lab and tell him that it would be in the Lab's best interests and Larry Roberts's best interests if the Director of Lincoln Lab would convince Larry to come down and take this job?' Charlie smiled and said, 'Sure,' picked up the phone, called Gerry Dineen at Lincoln Lab. I heard Charlie's end of the conversation – it was a short conversation – and two weeks later Larry accepted the job.[24]

Roberts was twenty-nine when he arrived in Washington. Within weeks his obsessive powers of concentration had become an office legend. It is said, for example, that he did an intensive study of the Byzantine, five-ring topography of the Pentagon building, working out with a stopwatch the fastest routes from one office to another.

His view of the basic topology of the network was similarly clear. One of his initial sketches[25] shows a minimal system linking the University of California at Santa Barbara (UCLA), the Stanford Research Institute (SRI), the University of California at Berkeley and the University of Utah.[26] A later development shows this core linked to System Development Corporation (SDC) at Santa Monica, the Universities of Michigan and Illinois, while a third extension takes in Project Mac[27] at MIT, Harvard and Carnegie-Mellon University (CMU) in Pittsburgh.

The nodes of the network were the large time-shared computers at these sites – called 'hosts' in network-speak. ARPA's initial idea was to connect all these machines to one another directly over dial-up telephone lines. The networking functions would be handled by the host computers at each site, which meant that each machine would, in effect, perform two tasks – its normal research-oriented number-crunching and the new task of routing messages.

This idea, however, did not appeal to the guys responsible for running the host machines. Nobody knew how onerous the routing task would be, and they were not keen to allocate scarce computing resources to servicing a network about which few of them were enthusiastic. At a meeting called in early 1967 at Ann Arbor to discuss the network, therefore, Roberts and his colleagues found themselves facing a wall of hostility from researchers on the ground.

Aside from the animus deriving from researchers protecting their own turf, there was also unease at the meeting about the whole everything-connected-to-everything-else concept as articulated by Roberts. In the end, the deadlock was broken by Wesley Clark – the guy who had first introduced Licklider to interactive computing almost a decade earlier. He listened quietly to the discussion and said little. But towards the end, just before they broke up, the penny dropped in his head. 'I just suddenly realised', he recalled, 'that they had an n-squared [n^2] interaction problem within computer nodes, and that that was the wrong way to go about it.'[28]

What Clark was referring to was a well-known problem – the rapidity with which the complexity of a densely interconnected network escalates as the number of its nodes increases. If you have a three-node network in which every node has a direct connection to every other node, then the number of connections needed is three. If you have four nodes, then the number of connections is six. If you have five, it's ten. If you have ten nodes, it's forty-five. And so on. In fact if n is the number of nodes, then the formula for the number of connections required to link them all is given by the formula:

$$\text{Number of connections} = \frac{n\,(n-1)}{2}$$

Which is just a mathematical way of saying that the connectivity increases as the square of the number of nodes. This is what Clark meant by the 'n-squared interaction problem'.

Clark reasoned that a network configured this way would be a disaster. It would be hard to fund and even harder to maintain and control. His great idea was simply to define the network as something self-contained without those n troublesome nodes. 'Leave them out; put them outside the network,' he reasoned. 'They weren't part of the network, the network was everything from there in, and that should be around a single, common message-handling instrument of some kind, so that the design of that instrument and all of the lines were under central control – ARPA central control.'[29] What Clark was saying, in effect, was that Roberts & Co. had conceived the network inside out. He suggested that they leave the host mainframes out of the picture as much as possible and instead insert a small computer between each host and the network of transmission lines. The small computer could then be dedicated solely to the message-passing process.

It was a breathtakingly simple idea – a subnetwork with small, identical nodes, all interconnected. Such a system would make few demands on the host computers – and would therefore make life easier for the people running them. It would mean that the specialised network computers could handle all the routing and could all speak the same language. And for ARPA it had the added attraction that the entire communications system could be under its direct control.

There is some disagreement about how Clark's proposed topology was firmed up into an actual design. Hafner and Lyon claim that Larry Roberts wrote a memorandum outlining Clark's idea when he returned to Washington.[30] Peter Salus maintains that the credit belongs to Elmer Shapiro, who wrote a report entitled 'The Study of Computer Network Design Parameters' at SRI under an ARPA contract in 1968.[31] But, however the design was elaborated, there is general agreement that the intermediate computers which would control the network were christened 'interface message processors' or IMPs. The IMPs would make the network function by sending and receiving data, checking for errors, routing messages and verifying that messages had reached their intended destinations.

The revised network idea was again put to the ARPA research community at a conference in Gatlinburg, Tennessee held towards the end of 1967. It was at this conference that the term 'ARPANET' was heard for the first time. The new design was received with more enthusiasm by those whom it was supposed to link. The guys who ran the host machines liked the fact that their precious machines would not have to devote scarce resources to routing and other mundane tasks – they would just have to communicate with a single IMP. And being computer folk, they probably also liked the idea of having another fancy machine on the premises, possibly because they did not yet appreciate the extent to which the IMPs would be under ARPA control.

In this way the basic topology of the network emerged. What nobody yet knew, however, was how the thing could actually be made to work. The only idea Roberts had was that the IMPs would be linked by ordinary 2,000 bits-per-second dial-up telephone lines (along the lines of the 1965 Roberts–Marill experiment which Bob Taylor had persuaded him to conduct), but beyond that they were groping in the dark.

Fortunately, there was someone in the audience at Gatlinburg who had the answer. He was a British researcher called Roger Scantlebury, and he came bearing news of what he and his colleagues at the UK National Physical Laboratory had been doing in the message-switching field. They had, in fact, already cracked the problem. But perhaps the most important news he conveyed was that the key to a viable network had been gathering dust on Pentagon desks for several years. It was contained in some obscure reports by a guy nobody at ARPA had ever heard of. His name was Paul Baran.

6:
Hot potatoes

Let them not make me a stone and let them not spill me,
Otherwise kill me.

                    Louis MacNeice, 'Prayer before Birth', 1944

The phrase 'father of the Internet' has become so debased
with over-use as to be almost meaningless, but nobody
has a stronger claim to it than Paul Baran. A laconic, understated,
clear-thinking, good-humoured, unpretentious engineer, he would
be the first to decry such a vulgar plaudit, but the fact remains that
he got to the key ideas first.

He was born in 1926 in Poland, and arrived at the age of two in
the United States. The Barans settled in Philadelphia, where his
father ran a small grocery store. Paul went to Drexel Institute of
Technology (later Drexel University), a no-bullshit engineering
school dedicated to the proposition that agility at mental arithmetic was next to godliness. After graduating with a degree in
electrical engineering in 1949 he saw a job advertisement in a local
paper for a post in the Eckert-Maunchly company, one of the early
computer manufacturers.[1] Baran applied, got the job and began his
working career testing racks of electronic components.

His initial encounters with the world of computing were not, to
say the least, encouraging. Indeed, they would make the basis of a
passable Charlie Chaplin movie. Computers had not figured on the
Drexel curriculum when he'd been an undergraduate, and, reckon-

ing that he ought to find out about the things, Baran enrolled for a course at the University of Pennsylvania. He was a week late in registering and missed the first lecture. 'Nothing much is usually covered in the first session anyway,' he recalled, 'so I felt okay to show up for the second lecture in Boolean algebra. Big mistake. The instructor went up to the blackboard and wrote 1+1=0. I looked around the room waiting for someone to correct his atrocious arithmetic. No one did. So I figured out that I may be missing something here, and didn't go back.'[2]

Back at Eckert-Maunchly there was precious little to boost his confidence in computing either. The company hit financial trouble after one of its main backers was killed in a plane crash. Baran decided the time had come to quit after a lunchtime conversation with some of his workmates. 'I was quietly munching on a (salami with cucumber on pumpernickel) sandwich,' he recalled,

> taking it all in as the older and wiser engineers discussed the company and its future. I think it was Jerry Smollier 'the power supply guy' who thought that the company had orders for three machines. 'Okay,' someone said, 'let's assume that we get orders for another three – six computers. Okay, let's double it again. What sort of business is it if your market is only a dozen machines?'[3]

Having left the computer business for dead, Baran then got a job with Rosen Engineering Products, a dead-beat distributor of TV sets whose bread-and-butter revenue came from repairing police and taxicab radios. The company had recently diversified into the manufacture of tele-metering systems – that is devices for recording remotely conducted measurements on to magnetic tape[4] – and Baran was put to work designing circuitry for correcting errors when recording FM radio signals. Among his discoveries was a method of eliminating tape noise caused by the deposit of tiny conical clumps of magnetic material. The solution was to scrape off the cones with a knife during manufacture to create what was called 'deteatified tape' until some humourless marketing executive tumbled to the mammarian connotation and had it relabelled 'instrumentation tape'.

To the astonishment of the company's management, the telemetry business took off like a rocket. The Air Force commissioned Rosen Engineering to provide the tele-metering for the first test of its Matador long-distance missile. Baran and his colleagues moved to Cape Canaveral in Florida, lived in a beach house, worked like crazy and produced test equipment that worked perfectly. He returned home with a pile of orders, and delightedly told the company's sole proprietor, Raymond Rosen Snr, of his achievement. Baran's account of the ensuing conversation goes like this.

'So, boychuck, where are we going to get all the people we need to do the contracts?'
'We have to hire them.'
'So, why will they work here?'
'We will have to pay more than they are now making.'
'So,' snorts Rosen, 'the guys fixing police and taxicab radios down the hall see this and come running to me crying for more money. If I give it to them I won't be able to compete.'
'Gee,' responds Baran, 'but the opportunity is huge. Telemetering and remote control are going to become big businesses one day.'
'Telemetering, schmetering – a flash in the pan! It will go away tomorrow. Now taxicabs and police radios I know. They are going to be around a long, long time after all this government electronics stuff disappears . . .'

Enervated by his experience of business. Baran spent a year doing consulting work before marrying a Californian woman and moving to the West Coast. He got a job with the Hughes Aircraft Corporation working initially on radar data-processing systems and later as a systems engineer specifying transistor subsystems. Since he hadn't a clue how a transistor worked, he enrolled in evening classes at UCLA and got a Master's degree. Baran then signed up for a PhD programme. By this stage he had landed a job at the RAND Corporation in Santa Monica and he found the commuting between RAND and UCLA too onerous. He claims that the final straw was arriving at UCLA one night and not being able to find a place to park. 'At that instant,' he said, 'I concluded that it was God's will that I should discontinue school. Why else would He have found it necessary to fill up all the parking lots at that exact instant?' Which is how Paul Baran came to abandon academia and embark upon the project which was to change the world.

RAND was a Californian think-tank founded in 1946 to provide intellectual muscle for American nuclear planners. Its roots lay in the intensive use of operations research[5] during the Second World War and it prided itself as an institution capable of applying 'rational' analysis to what most people regarded as unthinkable scenarios involving the incineration of millions of people and the wholesale pollution of the planet by nuclear warfare. It called its work 'systems analysis' – a discipline once lampooned by a colleague of mine as 'operations research plus megalomania'. Many of the so-called strategic doctrines which underpinned the American stance during the Cold War originated at RAND.

As an organisation, RAND has had an undeservedly bad press. It has suffered from being tarred with Dr Strangelove's brush. People have unloaded on it their distaste for the inhuman stand-off of the Cold War. But in fact it was a surprisingly enlightened place, resourced and managed on liberal lines conducive to creative work. It was funded mainly by the US Air Force, but in an unusual way. The Air Force awarded the Corporation a grant once a year but gave it considerable freedom in how it spent the money. Every week, RAND management circulated the letters which had come in from the Air Force and other federal agencies requesting help on various projects. If a staffer was interested or could make time then he or she could sign up for the project. If no one volunteered, RAND would respond with a form letter thanking the agency for its request but saying regretfully that the subject had been judged 'inappropriate' for RAND.

At the same time, however, the Corporation conducted a number of major, formal projects, generally in areas of acute interest to its prime sponsor. These 'support projects' were regarded by the management as essential to preserve the Air Force's goodwill and thereby allow other parts of the organisation more freedom to do what they wanted. It was a remarkably open, relaxed and productive regime which treated researchers like responsible adults – and, not surprisingly, elicited highly responsible and committed work from them. 'If you were able to earn a level of credibility from your colleagues,' said Baran, 'you could, to a major degree, work on what you wanted to. The subject areas chosen were those to which you thought you could make the greatest contribution.'[6]

Baran joined RAND in 1959 at the age of thirty. Significantly, he was recruited by the computer department, not the communications department. His arrival coincided with a critical period in the Cold War. 'Both the US and USSR were building hair-trigger nuclear ballistic missile systems,' he recalled.

> The early missile-control systems were not physically robust. Thus, there was a dangerous temptation for either party to misunderstand the actions of the other and fire first. If the strategic weapons command and control systems could be more survivable, then the country's retaliatory capability could better allow it to withstand an attack and still function; a more stable position. But this was not a wholly feasible concept because long-distance communications networks at that time were extremely vulnerable and not able to survive attack. That was the issue. Here a most dangerous situation was created by the lack of a survivable communication system. That, in brief, was my interest in the challenge of building more survivable networks.[7]

This had long been a hot topic within the Corporation, and its researchers had been pondering it before Baran's arrival. In a report issued in 1960, RAND estimated the cost of a nation-wide, nuclear-hardened, buried cable network at $2.4 billion.[8] The new recruit was set to work with the brief of determining whether it was possible to do better for less. Baran's opening conjecture was that it should be possible to build a survivable system. In just under two years, he devised a workable design. Then he spent the next three painstakingly answering those who claimed it could not be done.

At its crudest level 'command and control' in nuclear war can be boiled down to this: *command* means being able to issue the instruction to 'fire' missiles, and *control* means being able to say 'cease firing'. Baran's first stab at the problem was to ask: what is the minimum requirement for a survivable communications system?

He concluded that it was a single teletype channel enabling the President of the United States to say 'You are authorised to fire your weapons' or 'Hold your fire'. Baran resurrected an idea which had been originally proposed by Frank Collbohm, President of RAND, who had suggested using the extensive US network of AM[9] radio stations – the ones which generally broadcast wall-to-wall country and western music – to relay messages from one station to another.

Baran showed how, with the addition of a little digital logic at each of the broadcast 'nodes', it would be possible to get a simple message across country in almost any circumstances. The strategy would be to flood the network with the message. No routing was required other than a procedure which ceased transmission of the message once copies of it started bouncing back. Baran's calculations indicated that even with widespread damage to metropolitan radio stations, the overlapping nature of the US commercial broadcasting system would allow the message to diffuse rapidly through the surviving stations on the network. This – as George Dyson has pointed out – had the interesting implication that American country music would have survived even a worst-case Soviet attack.[10]

This idea was duly presented to the US military. The top brass were distinctly underwhelmed. Such a primitive system might be okay for the President, they observed, but *they* needed much higher bandwidth. (It is an ancient law of military life that high-ranking officers need secure, high-capacity communications channels not only for communicating with their superiors but also for the more subtle nuances required for messages to wives and mistresses, champagne importers and so on.)

Okay, said Baran, and went back to his drawing board, determined, as he put it, 'to give them so damn much communication capacity they won't know what in hell to do with it all'. He started from three very simple propositions: one, avoid centralisation like the plague – because any centralised system can be disabled by a single well-aimed strike; two, build a *distributed* network of nodes, each connected to its neighbours; and three, build in a significant amount of *redundancy* in the interconnections (that is, have more connections than you strictly need for normal communications).

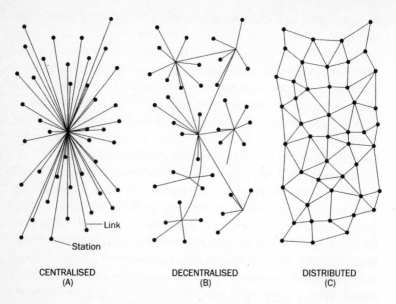

CENTRALISED   DECENTRALISED   DISTRIBUTED
     (A)   (B)   (C)

These ideas are illustrated in one of his earliest drawings (see figure).[11] The aim was to avoid both (A) and (B) and go for agenuine, distributed 'net' (C). Given that aim, the central question was how much redundancy was necessary to ensure survivability. Baran did some computer simulations which showed the answer was less than expected. In fact, he calculated that if each node were connected to three others then the resulting network would be acceptably resilient. This was, he recalled, 'a most fortunate finding because it means that we would not need to buy a huge amount of redundancy to build survivable networks – just a redundancy level of maybe three or four would permit almost as robust a network as the theoretical limit. A pleasant, unexpected result.'[12]

Plumping for a distributed network had one inescapable consequence, namely that *the signals which traversed the net would have to be digital ones*. The reason was that it would be impossible to achieve the level of redundant connections required with the conventional (analog) telephone system because signal quality deteriorated rapidly when several links were connected in tandem.

At each switched link the signal would become slightly more degraded.

It's a bit like what happens with bootleg audio recordings. Someone buys a cassette of, say, a Bruce Springsteen album and lends it to a friend, who makes a copy (Copy A) using the tape-to-tape facility in his ghettoblaster. He then lends the illegal copy to another friend, who likewise makes her own copy (Copy B). The process goes on until, by the time you get to Copy E, the audio is so degraded that someone starts muttering about getting their money back.

Digital signals, however, do not degrade in the same way – and, if they do, error-correcting procedures can be used to correct the distortions. Digital signals are, after all, only sequences of ones and zeroes, and there are simple techniques for checking whether a given sequence has been transmitted correctly. If it hasn't, then a request for retransmission of the damaged sequence can be issued – and if necessary repeated until the correct sequence is received. Digital technology could thus allow the creation of net-works which communicated data accurately while being densely enough interconnected to provide the redundancy needed to ensure resilience in the face of attack. So, for Baran's distributed net to work, each node had to be a message-switching *digital* computer.

Puzzled by all this analog vs digital stuff? Read on.

Analog communications

When you talk, your vocal chords create vibrations or pressure waves which travel through the air. When you speak into an ordinary telephone, the diaphragm in the handset microphone converts your speech into electrical waves which are analogous to (analogs of) the original pressure waves. The electrical waves then travel along the wires of the telephone system until they reach the receiving handset, where they cause another diaphragm to vibrate and produce pressure waves which can then be heard by another person.

In idealised form, an analog wave looks like this. Of course real speech waveforms (and their corresponding electrical analogs) are more complex, but the principle is the same:

You can get a good image of a wave and how it travels by imagining two children holding a skipping rope. One gives a sharp flick and produces a wave which travels along the rope to the other end.

The trouble with analog signals of this kind is that they degrade easily. They are drastically attenuated (weakened) by distance, for example. (Imagine a skipping rope that was a mile long: the wave would have died out long before it reached the other end.) And they are easily corrupted by interference, noise and switching, so that the received signal bears a poor relationship to the original.

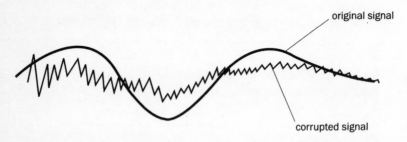

original signal

corrupted signal

Digital communications

Digital communications work by taking analog signals, converting them into sequences of ones and zeroes which are then transmitted through a channel (telephone line, wireless link, whatever) to a receiver which then reconstructs the analog signal into something that can be heard by the human ear.

This is done by *sampling* the analog wave, converting the samples into numerical values and thence into ones and zeroes, like this.

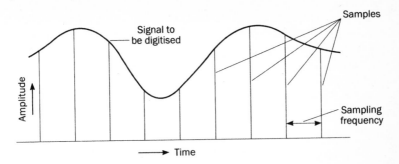

At the receiving end, the process is reversed to reconstruct an image of the original analog signal.

You can see that the quality of the digital representation of the signal is critically dependent on how often it is sampled. Think of trying to drive a car wearing a set of goggles with shutters which open and close at regular intervals. If the shutters open and close sixty times a minute then you will probably be able to drive reasonably safely – unless traffic conditions are changing very rapidly. But if the shutters operate only once every thirty seconds then you will probably find it impossible to drive.

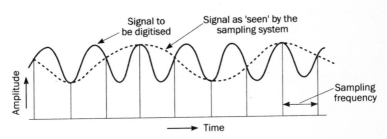

Here the signal frequency is too high for the sampling frequency, resulting in the wrong signal being seen by the sampling system.

Much the same applies to sampling rates for digital signals. The more frequently you sample, the more accurate the representation of the analog signal. But here, as everywhere else, there is no such thing as a free lunch, for the more frequently you sample the more digital data is generated – and the longer the bit-stream which has to be transmitted.

The great thing about digital signals though is that even if they become corrupted in transmission, the errors can be detected and corrected. There are various error-correction methods available. For example, the transmitter can add a special bit (called a parity bit) at the end of each group of digits, depending on whether the group has an even or an odd number of ones in it. On receiving the group, the receiver checks to see whether the sequence of bits has the right number of ones. If not, it requests retransmission of the group. This is, of course, an incredibly tedious business even to describe. But it's the kind of things computers can do a million times a second without batting a virtual eye.

Baran's next great idea was to break the messages into small, regularly sized pieces. By dividing them into what he called 'message-blocks' it would be possible to flood the network with data segments, all travelling along different paths towards their destination(s). No node on the network would send a large message in one piece – any more than one would contemplate moving a large house in one piece from New York to San Francisco. The obvious way to do it is to disassemble the structure, load segments on to trucks and then dispatch them over the interstate highway network. The trucks may go by different routes – some through the Deep South, perhaps, others via Kansas or Chicago. But eventually, barring accidents, they will all end up at the appointed address in San Francisco. They may not arrive in the order in which they departed, but provided all the components are clearly labelled the various parts of the house can be collected together and reassembled.

Baran's resilient network did something similar with digital messages. It consisted of a network of computerised switches which passed message-blocks from one to another without any centralised control. But, without central control, how would a node know where to send each message-block it handled? Baran's solution to this problem was to invent a procedure to enable each node to choose the best route for the next stage of the block's journey across the net. He proposed an algorithm – which is just a fancy name for a computerised procedure – which operated on the principle of what he called 'hot potato routing': it was a rapid store-and-forward routine which took in message-blocks and passed them on as quickly as possible, rather like a group of people passing hot potatoes from one to another, with everyone trying to ensure that they handled the potato as little as possible. The core of Baran's algorithm was a constantly updated table containing data about how many 'hops' were needed for a message to reach every other node in the system. The table would indicate the best routes to take at any moment in time and was regularly updated by news about the state of neighbouring nodes. If one node went down, then the table was updated to reflect the change and message-blocks were then switched to other routes.

These ideas were worked out not in the peace and quiet of a laboratory, but in the incongruous setting of an airline seat. Shortly after he joined RAND Baran was nominated to serve on a high-level committee set up by the Pentagon to choose a communications system for handling Department of Defense records. The committee met every other week in Washington, so Baran was obliged to spend one week at his home in California and the next week in the capital. Though it was in one sense a chore (he was effectively standing in for the Director of RAND), Baran's membership of the committee proved fortuitous on two counts.

In the first place, it gave him a licence to be inquisitive about what was then regarded as state-of-the-art in switched communications networks. 'I was able to ask some really dumb questions,' he recalls. 'Dumb questions are only allowed without giving offense if you are a child or a member of a distinguished evaluation committee. "Why are these things so big?" "Why do these switching systems require rooms and rooms of stuff?" '[13] What prompted these inquiries was the fact that all the systems proposed to the committee required massive mainframe computer systems as the nodes. This is what made any ambitious switched network seem prohibitively expensive (at the time, remember, mainframe computers cost millions of dollars and required large retinues of keepers).

Baran pressed on to the heart of the problem: why did people think these massive nodes were necessary? The answer, he discovered, was that they kept elaborate records of all the transactions they had handled. 'It seems', he concluded, 'that the real reason every communications center office was built with that burdensome capacity was to be able to prove that lost traffic was someone else's fault.' Baran realised that the expensive mainframes were essentially blame-avoidance systems! The corollary was that if you weren't bothered about covering your ass you could get by with much simpler (and cheaper) machines.

The other benefit Baran got from being on the committee was lots of peace and quiet. Because of the heavy travelling involved he had plenty of uninterrupted time on aeroplanes, flying backwards and forwards between California and Washington. 'The airplane seat is a wonderful place to spread out and work uninterrupted,' he recalled.

> In those days RAND allowed its staff to travel first class if the trip was over two hours. This gave me the equivalent of one day per week to work in a comfortable setting. Almost all my writing was done either in airport waiting rooms or in airplanes. Today the difference in price between coach [economy] and first class is so great that few organizations can justify first class travel any longer. In the old days the price differential was small. It's a pity because present day steerage type, crammed together air travel poses an invisible impediment to economic productivity by the business traveller.[14]

Baran's grand design for a survivable communications network took surprisingly little time to germinate. The whole concept was essentially complete by 1962. It consisted of a network of 1,024 switching nodes which sent and received message-blocks via low-power microwave transmitters mounted on small towers. Each node was a relatively modest computer. By keeping the distance between towers short – of the order of about twenty miles – only short and inexpensive structures were needed. The transmitter/receiver units were 'about the size of a shoe-box', constructed from transistorised components and designed to plug directly into the centre of a moulded-foam-plastic dish-shaped antenna. Even for the time it wasn't state-of-the-art technology; in fact it was conceived as an inexpensive lash-up made from mostly standard components. Baran himself referred to it as a 'poor boy' system.

Since destruction of the national electricity grid was an assumed consequence of a nuclear exchange, the transmitter/receivers were to be powered by small generators fuelled by liquid petroleum from a 200-gallon tank buried in the ground beneath each tower. As each relay-tower would consume only fifty watts of electricity, this local supply of fuel would last for at least three months after failure of the grid.

From the outset, Baran's system was designed to carry 'secure' (that is, encrypted) voice as well as digital data. It also had formidable capacity. As we have seen, when his initial ideas for using AM stations were rejected by the Air Force, he had sworn to give generals more bandwidth than they would know how to handle. And he was as good as his word: each node in the system could handle 128 cryptographically secure telephone subscribers, together with 866 other simultaneous subscribers using computers and other digital devices. On the data side it was designed to handle virtually every communication requirement likely to be thrown at it – from primitive, low-speed teletype/telex signals to digital data rates up to 19,800 bits/sec.[15] That was in 1962.

Baran published his ideas about a survivable network in the open literature in 1964.[16] In retrospect, the freedom to publish secrets of this order seems astonishing. Just think: here were the results of advanced research on one of the most sensitive areas in US nuclear strategy – and they were placed in the public domain. The US, in other words, just gave them away! They weren't even patented. But in the topsy-turvy context of the nuclear stand-off, it made perfect sense to do this. 'We felt', said Baran, 'that it properly belonged in the public domain.'

Not only would the US be safer with a survivable command and control system, the US would be even safer if the USSR also had a survivable command and control system as well! There was never any desire for classification of this work. Classification tended to be limited to analysis of vulnerabilities; how many sites would be damaged, what our assumptions were of the other guy's capabilities,

versus our capabilities. They are about the only things that ever got
classified.[17]

It is one thing to be free to publish one's ideas, but quite another to
persuade others to take them seriously. There was a saying in RAND
that 'It takes thirty to sixty briefings to sell an idea', and the
survivable network proved no exception. Baran spent the best part
of three years and compiled a mountain of paper trying to persuade
people that his system would work.

Some of the objections were technically informed and reason-
able. How would the cryptography work? How reliable were his
estimates of the level of redundancy needed to ensure survivability?
How would the system behave under heavy loading? Under what
conditions would it become saturated and vulnerable to lock-up?
One by one, Baran and his colleagues faced up to these intellectual
challenges and dealt with them in meticulous detail, in the process
compiling thirteen detailed technical reports.

All of which was par for the course. To be worthwhile, any
technological scheme has to be able to survive vigorous critical
examination. Much more disconcerting to Baran, however, was the
discovery that the main opposition to his ideas came from people
who seemed immune to rational discussion. What made it even
more peculiar was that they were the guys who ought to have
known most about communications – the men from the giant
telephone company AT&T. And the most outspoken of the 'it ain't
gonna work' school, Baran discovered, were the most senior AT&T
technical management people. The company at that time held a
total monopoly of all long-distance communications within the
US. And AT&T headquarters people insisted that his design was
not feasible.[18] The discussions between Baran and his adversaries
were generally courteous. 'One helpful fact', he said, 'was that
AT&T was composed of gentlemen. Talk politely to them and they
would invariably talk politely back to you.' But behind the urbane
façade the opposition was implacable and unyielding.

There were a number of corporate reasons for AT&T's obduracy.
For one thing, accepting Baran's ideas would have been tantamount

to conceding the premise behind all his work – that there was a vulnerability problem with the AT&T network. This was something the company had always denied. All military telephone traffic in the US went out on AT&T lines. The company's proudest boast was that it ran the best telephone system in the world. How could one change the basic architecture of such a superb system without undermining that claim?

There were also some technical reasons for going into denial. The AT&T system was a monolithic, analog one, and everything added to it had to work with the older stuff. 'At that time,' said Baran, 'there was simply no way to bring radical new technology into the plant and have it coexist gracefully with the old.'

But the most important reason for opposition was the simplest one, namely that AT&T's most senior engineers just did not understand what digital technology and computer-switching would mean for telecommunications in the coming decades. Their views were once memorably summarised in an exasperated outburst from AT&T's Jack Osterman after a long discussion with Baran. 'First,' he said, 'it can't possibly work, and if it did, damned if we are going to allow the creation of a competitor to ourselves.'[19]

Back at RAND however, Baran's colleagues and superiors were satisfied that his work was sound. In 1965 the Corporation formally recommended to the Air Force that it should construct an experimental version of his network. 'In our view,' wrote the President in his covering letter,

> it is possible to build a new large common-user communication network able to withstand heavy damage. Such a network can provide a major step forward in Air Force communication capability . . . Although the system is a marked departure from the existing communication system, we believe that after the R and D is proved out the new system can be integrated with the present one on an incremental basis.[20]

The Air Force staff accepted the proposal enthusiastically. Following normal procedures they set up an independent evaluation of it which was run by the MITRE Corporation on behalf of the Air Force System Command. The MITRE report, in its turn, was exceedingly positive and recommended that the project should proceed. It looked like plain sailing all the way.

And yet it never came about. The reason was that it ran into the bureaucratic equivalent of a brick wall. Since 1949, successive US administrations had been trying to bring about the unification of the three branches of the armed services. Because of inter-service rivalries, it proved to be slow work. But in 1960 President Kennedy had installed as his Secretary of Defense a guy called Robert Macnamara, a steely managerialist who had previously run the Ford Motor Corporation. The profligacy of inter-service rivalry offended Macnamara's sense of bureaucratic order and he set about knocking heads together.

By the time Baran's project reached the Pentagon, Macnamara's attempts at rationalisation had produced a new organisation called the Defense Communications Agency. This had been given overall responsibility for all long-distance communications for the three armed services. In the manner of these things, it was constituted on a modified Noah's Ark principle – a few of everything. It was headed, for example, by an Air Force general, an Army general and an admiral. More significantly, it was run mainly by ex-AT&T people who had zero or near-zero exposure to the new technology. 'If you were to talk [to them] about digital operation,' mused Baran, 'they would probably think it had something to do with using your fingers to press buttons.'

The problem as Baran saw it was that, if his RAND project were to go ahead, these boobies would be charged with building it. And that, he thought, would be tantamount to giving his design the kiss of death. In desperation, he consulted Frank Elldridge, a personal friend who worked in the office of the Assistant Secretary of Defense and was a key person in communications funding decisions. 'Frank and I agonised,' recalled Baran.

We agreed that DCA had the charter. The legal determination had been made. We also agreed that the present DCA wasn't up to the task. I felt that they could be almost guaranteed to botch the job since they had no understanding for digital technology, nor for leading edge high technology development. Further, they lacked enthusiasm. Sometimes, if a manager doesn't have the staff but has the drive and smarts to assemble the right team, one could justify taking a chance. But lacking skills, competence and motivation meant backing a sure loser.[21]

Elldridge and Baran reluctantly concluded that the project should not proceed because they thought the DCA would screw it up. 'The risk was compounded', said Baran, 'because we both knew that if the project turned into a botch, it would be extremely difficult to get it going again. Detractors would have proof that it couldn't be done.' Baran didn't want his baby to be strangled at birth by an incompetent and malevolent midwife. So he aborted it himself. Looking back at it now, the whole thing seems crazy. But it happened.

7:
Hindsight

Hindsight is always twenty–twenty.

Billy Wilder.

With the omniscience of hindsight, it's easy to ridicule the AT&T personnel who were so resistant to Baran's ideas. Every telephone system in the modern world now runs on digital technology. How *could* they have been so dumb? Surely their obtuseness must have had something to do with the fact that they worked for a semi-monopoly which had a massive investment in the technology which Baran's ideas threatened with obsolescence?

No doubt. But the history of science and technology is littered with stories of developments which look irresistible or inevitable to us and yet were ignored or vigorously resisted at the time. What we forget is that this perspective is a by-product of the way intellectual history is usually written. We tend to portray it as a record of continuous, cumulative progress. But doing so obscures the fact that at any moment the future is unknowable. What seems obvious in restrospect may not have been at all obvious at the time.

The most celebrated exponent of this insight is Thomas Kuhn, the great historian of science whose book *The Structure of Scientific Revolutions*[1] radically altered the way we think about intellectual progress. He started from the view that one should give scientists of the past the benefit of the doubt rather than viewing them with the condescension endowed by hindsight. One should always assume, for instance, that they were intelligent beings doing their best to make sense of complex phenomena.

Kuhn argued that scientists (and by inference all members of scholarly disciplines) work within an established theoretical framework which he called a 'disciplinary matrix' or 'paradigm'. This represents the consensus about the discipline's knowledge at any given point in time – the theoretical knowledge embodied in textbooks, for example, and the craft knowledge imparted in laboratory tutorials. The paradigm defines what counts as an explanation, what are acceptable methods of inquiry, what constitutes evidence, and so on.

No paradigm is perfect, however, because there will always be discrepancies between what is observed in the real world and what can be explained by the paradigm. These discrepancies or anomalies are 'puzzles', and most of the time scientists spend their time trying to solve them in a process which Kuhn called *normal science*. The paradigm thus defines what constitutes interesting problems: they are, for example, the puzzles which graduate students are set for their doctoral research. Most of the time such puzzles can be solved, either by demonstrating that they are not really anomalies at all, or by minor elaborations of the pardigm to bring the anomalies within its explanatory reach.

However, some puzzles stubbornly elude resolution, and over time the list of these recalcitrant anomalies grows longer and more troubling until, in the eyes of some practitioners, they threaten the very paradigm itself. At this point the discipline enters a period of turmoil during which some scientists actively espouse a new paradigm, while the majority continue to support the old one. The crisis is heightened because the two paradigms are, as Kuhn put it, 'incommensurable'. That is to say, there exists no common, agreed-upon frame of reference within which the merits of the old and new paradigms can be objectively discussed. Kuhn likened these periods of intellectual ferment to the turmoil which precedes political revolutions. And he maintained that scientific paradigms, like political orders, are *overthrown*. The old paradigm is replaced by the new one and normal science is resumed using the replacement as the disciplinary matrix.

This interpretation of the way scientific disciplines evolve has had a radical effect on the way we think about knowledge. From our point of view, however, the most important point about it is the significance it assigns to the paradigm. An intellectual community does not lightly adopt a paradigm – indeed that is why a revolutionary challenge to one is so fiercely resisted. A paradigm is a powerful theoretical and methodological framework which defines the working lives of thousands of intelligent and disciplined minds. And paradigms do not attract the loyalty of such minds unless they 'work'. One of the first things a graduate student learns is that if there is a discrepancy between the paradigm and what he or she has discovered, then the automatic assumption is that the paradigm is right and the student wrong. Just as a good workman never blames his tools, so the diligent student never blames his paradigm.

Science is different from technology because one is about *knowing* and the other is about *doing* – about making things that work. But, although the concerns of engineering disciplines may be more practical than those of their scientific counterparts, they are intellectual communities too, and Kuhn's model of how knowledge develops can be used to illuminate the process by which technological progress comes about.

The first thing to note is that there is such a thing as 'normal technology'. At any given time, in any given specialism, there will be a community of practitioners working with a shared set of intellectual tools and methodologies. 'Technological traditions of practice', wrote Edward Constant in his wonderful history of the turbojet revolution in aeronautics,

> comprise complex information physically embodied in a community of practitioners and in the hardware and software of which they are masters. Such traditions define an accepted mode of technical operation, the conventional system for accompanying a specified technical task. Such traditions encompass aspects of relevant scientific theory, engineering design formulae, accepted procedures and methods, specialised instrumentation and, often, elements of ideological rationale.[2]

There are, in other words, such things as *technological paradigms*, and they are determined by community practice. And just as normal science is a livelier business than Kuhn's term implies, so too is normal technology. It is a process of continual improvement and refinement. Most engineers have Henry Ford's dictum – 'everything can always be done faster and better' – engraved on their hearts. Technological systems are complex and have many components; research and development on the components continually throws up changes and improvements, which in turn give rise to developments in the nature and performance of the wider systems to which they belong. (Think, for example, of what has happened to computers and modems – or, for that matter, automobiles – in recent decades.) Change is thus incessant in normal technology – it just happens in an evolutionary, incremental way.

But revolutions also occur in technology as in science, and they are just as traumatic because communities do not lightly abandon their paradigms. Why? Because they have huge investments – organisational, personal, intellectual and occupational – in them. We have seen that what unhorses a scientific paradigm is the perception of significant anomalies between the physical world and what the paradigm can handle. The question is: what causes similar upheavals in the engineering world?

There are basically two kinds of situations which can lead to revolutionary change in technology. One is *functional failure* – that is, an occasion where the prevailing paradigm fails spectacularly, and where evidence of its failure is unequivocal and unchallengeable. A bridge collapses; a plane falls out of the sky; an electricity grid fails; a nuclear reactor goes into meltdown. Actual failures of this kind, however, are comparatively (and mercifully) rare. What is more common is the realisation that the existing paradigm fails (or is likely to fail) in what Kuhn called 'new or most stringent conditions'. For example, as First World War combat aircraft flew higher and higher in the quest for aerial superiority, their non-supercharged piston engines faltered as the oxygen in the air diminished at high altitudes.[3]

Although functional failure is a powerful force driving technological progress, it cannot be the only one because revolutionary technologies often appear long before conventional ones are seen to be in trouble. What's happening in such cases, Edward Constant argues, is not that current technology fails in any absolute or objective sense, but that 'assumptions derived from science indicate either that under some future conditions the conventional system will fail (or function badly) or that a radically different system will do a better job'.[4] Constant coined the phrase 'presumptive anomaly' to describe this phenomenon. It's an ugly phrase, but I think it offers us a way of understanding why the folk at AT&T thought that Paul Baran, with his high-falutin' theories about resilient, digital networks, was out to lunch.

To see why, we have to put ourselves in their shoes. The first thing to remember is that AT&T was a *telephone* company. Its business was to provide wires along which voice messages travelled, and switches (exchanges) to enable subscribers to make connections. Secondly, 'Ma Bell', as the company was called, enjoyed a special position in American corporate life as a monopoly – albeit a regulated one. In 1907 Theodore Vail, then President of AT&T, had argued that the telephone, by its nature, would operate most efficiently as a monopoly providing universal service. Government regulation, he maintained, 'provided it is independent, intelligent, considerate, thorough and just',[5] would be an appropriate and acceptable substitute for the competitive market-place.

For over forty years, successive US governments bought this argument, and AT&T thrived as a result, becoming one of the most powerful corporations in the United States. It maintained formidable research and development establishments. Bell Labs, for example, was the place where several of the defining technologies of the twentieth century were invented, among them the transistor, the UNIX operating system, the laser, the C programming language and much else besides. And the creation of an international analog telephone system was a great technological achievement. It involved massive investment in infrastructure (lines, exchanges) and a fantastic maintenance effort. And it had a dramatic impact on society, determining among other things, the shape of cities and the locations where people could work.[6]

By any standard that one cares to mention therefore, the

company which had brought this about was a formidable outfit. But its technical culture reflected its technology. The AT&T network was an analog one, and its people were analog people. That's not to say they thought the technology was perfect – far from it. They knew, for example, that it was inherently inefficient in terms of its use of facilities. Analog communication was only possible when a complete, dedicated circuit had been established between sender and receiver. Indeed, establishing such connections was the sole function of the telephone exchange, which in essence was just a large bank of (physical) switches. The circuit was maintained for as long as both parties desired, after which it was closed. But since a large part of any conversation consists of silence, this meant that the line carrying the conversation is effectively idle for much of the duration of the connection – a terrible waste of expensive resources.

The AT&T crowd devised a series of very ingenious methods for getting round the problem. Most of these involved a process called *multiplexing* – that is, mixing signals from different conversations in such a way that multiple conversations could be carried simultaneously on trunk lines between exchanges without interfering with one another. Other, equally smart ways of making the system more efficient (for example by the use of microwaves) were also developed in the kind of continuous development typical in 'normal' technology. To most observers in the late 1950s, therefore, the analog system would have seemed the epitome of high-tech, science-based progress.

With the omniscience of hindsight, of course, it looks quite different. *We* know, for example, that dedicated connections are a very wasteful way of providing channels for communication between computers, because digital signals (streams of ones and zeroes) are inherently 'bursty'. That is to say, they tend not to come in a steady stream, but in staccato bursts. When a burst is in progress, computers need a dedicated channel; but otherwise they might as well be disconnected from the system. If you have a switching technology fast enough to cope with this – quick enough to make a connection only when it's needed – then a digital system is inherently much more efficient in its use of infrastructure.

But in the late 1950s and early 1960s few telephone lines were used for carrying digital signals of any kind. The analog community had very little experience of such things. As far as they were concerned, they had an admirable system which worked just fine. It had not failed, and its reliability was increasing all the time. Indeed they were exceedingly proud of its dependability. So when Paul Baran knocked on their door telling them that (1) there were certain conditions (nuclear war) under which their cherished system could fail catastrophically and (2) the only way forward was to abandon the analog paradigm and adopt a new, unfamiliar one, they were simply incredulous.

What was happening, of course, was that Baran was presenting them with one of Edward Constant's 'presumptive anomalies' and they couldn't see it. In fact, their first response was to explain to Baran that his network ideas were 'bullshit' (Baran's word). When that failed, they set up a series of tutorials designed to explain their system to this troublesome and wrongheaded interloper. The tutorials lasted for several weeks. 'It took ninety-four separate speakers to describe the entire system,' Baran recalled, 'since no single individual seemed to know more than a part of the system. Probably their greatest disappointment was that after all this, they said, "Now do you see why it can't work?" And I said "No."'[7]

In fact, Ma Bell's scepticism about this new-fangled technology knew no bounds. In the autumn of 1972, after a network based on Baran's concepts was up and running, ARPA organised a massive demonstration of the network at a Washington conference. A young Harvard student called Bob Metcalfe (who later invented the Ethernet local area networking system) was detailed to give ten AT&T executives a virtual tour of the new network. In the middle of his demonstration the computers crashed. The telephone executives' first reaction was to laugh. 'I looked up in pain,' Metcalfe recalled, 'and caught them smiling,' delighted that this new-fangled digital technology was flaky. It confirmed for them that the analog way of doing things was here to stay. They didn't realise, Metcalfe said, that they were 'tangled up in the past'.[8]

The strangest thing of all is that long after the technical viability of the ARPANET had been proved beyond a shadow of a doubt, AT&T continued to be suspicious of it. Some time in late 1971 or early 1972 ARPA began to look for ways of divesting itself of operational responsibility for the network. Larry Roberts (then IPTO Director) approached AT&T and asked if they would like to take it over. After a committee comprised of senior managers *and* representatives from Bell Labs had considered the question for several months, the company declined. They finally concluded, Roberts said, that the new-fangled technology was incompatible with their precious system.[9] And, in a way, I suppose they were right.

8:
Packet post

The characteristic virtue of Englishmen is power of sustained practical activity . . .

R. H. Tawney, *The Acquisitive Society*, 1921

The Internet is as American as apple pie – everyone knows that, right down to the technophobe who sneers contemptuously at the idea of people 'surfing the Net'. And although the network has been around for a long time now, it's still the case that the values which dominate it are American ones. It is the US Constitution, for example, which allows non-US journalists and others to publish Websites which would be banned or censored if they were hosted in their home countries. But putting them on a server which is physically located in the US bestows upon them the formidable protection of the First Amendment, much to the chagrin of tyrants, secretive bureaucrats and paranoid executives everywhere.

It's also true that the Net is largely an American creation. But not entirely. For while the Americans running ARPA in 1967 possessed the vision, money and determination needed to turn Bob Taylor's dream into a physical reality of computers and communication lines, they lacked one vital ingredient. Since none of them had heard of Paul Baran they had no serious idea of how to make the system work. And it took an English outfit to tell them.

Around the time that Baran submitted his design to the USAF,

Donald Watts Davies, Head of Computer Science at Britain's National Physical Laboratory (NPL), was also pondering the deficiencies of the analog telephone system. Davies had two things in common with the folks at ARPA. The first was that he knew absolutely nothing of Baran's work at RAND; the second was that he was also interested in finding a way of enabling computers to communicate efficiently over telephone lines.

Donald Davies is one of those men who brings to mind Harry Truman's observation that 'you can accomplish anything you want in life provided you don't mind who gets the credit'.[1] A mild-mannered, quietly spoken, formidably intelligent man with a penchant for precision and understatement, he is one of the few computer specialists who has been elected a Fellow of the Royal Society – the highest distinction, short of a Nobel prize, available to a British scientist. As a child he was fascinated by telephone switching and used to design complicated logic circuits using electro-mechanical relays. Nowadays, his hobby is dissecting the design of the German Second World War 'Enigma', T52 and SZ42 encryption machines and the methods by which British cryptographers at Bletchley Park cracked the Germans' codes. In his retirement, this most cerebral of men runs a consultancy business specialising in computer security.

Davies began his professional life as a physicist, studying at Imperial College, London – the nearest thing Britain has to MIT – in the early years of the Second World War. Graduating with a First in 1943 he was immediately drafted to the 'Tube Alloys' project at Birmingham University – the code name for Britain's part of the Atomic Bomb project – where he worked for a time under the supervision of Klaus Fuchs, the celebrated physicist who was later exposed as a Soviet spy.

The Tube Alloys work gave Davies his first encounter with what were then known as 'computers' – that is, human beings employed solely to perform mathematical calculations. He was working on the design of the diffusion separator plant for Uranium 235, and this required intensive analysis which could only be done at the time by large numbers of people using mechanical calculators. At the end of the war, deciding that he needed to know more about numerical analysis, Davies returned to Imperial College to take a second degree – this time in mathematics. It was while he was there that his interest in computers in the modern sense was triggered – oddly enough, by a lecture on cybernetics given by Norbert Wiener. A subsequent talk at Imperial given by John Womersley of the NPL on the laboratory's exploratory work in computing clinched the matter. Davies went for an interview, was offered a job in this strange new field and accepted.

The National Physical Laboratory is a curious place, a uniquely British blend of bureaucracy and ingenuity. It was set up in 1900 in Teddington, an unfashionable town within easy commuting distance of London, as a response to German state funding of scientific research. It was the most extensive government laboratory in the UK and over the years carved out a high reputation in its assigned role of setting and maintaining physical standards for British industry. The head of the NPL was invariably a pillar of the British scientific establishment. At the time Davies joined the Laboratory's staff, for example, the Director was Sir Charles Galton Darwin, an eminent mathematician whose grandfather wrote *The Origin of Species* and changed for ever the way we think of ourselves.

Davies arrived at the NPL in August 1947 in the middle of a surge of interest in the strange new devices called digital computers. On both sides of the Atlantic, research teams were racing to construct what the popular press had already begun to call 'thinking machines'. In the US, massive research efforts were under way at Harvard (led by H. H. Aiken), at the University of Pennsylvania (Ekert and Maunchley) and at Princeton (John von Neumann). The post-war British government, conscious of the potential of this new technology and desperate not to be trumped by American mastery of it, had launched its own project to build an 'Automatic Computing Engine' (ACE), and Darwin and Womersley had lobbied successfully to have the project housed at the NPL. More significantly, Womersley had even lured Alan Turing – the mathematical genius who is the intellectual godfather of Artificial Intelligence and digitial computation generally – to NPL to run the project.

Turing had been at the Laboratory since October 1945. He had created an ambitious and ingenious design for ACE but had become disillusioned by the administrative inflexibility which had crippled the project almost from the outset. The NPL housed a large number of formidable scientific intellects, but it corralled them within the tidy organisational structures beloved of civil servants. Turing was maddened, for example, by the rigid demarcation which existed between 'theoretical' (that is, scientific) and 'technical' support staff. His experience of wartime code-breaking at Bletchley Park had proved the importance of having technicians as full members of project teams. He was himself accustomed to building and fiddling with electronic apparatus. Yet it took him over a year to persuade NPL management to establish an electronics facility, and when one was eventually created he was not consulted about the appointment of its first superintendent, who turned out to be someone largely uninterested in the ACE work.

The problem with ACE was not lack of money; on the contrary, the research was reasonably funded by the standards of the time. What was wrong was that the project had made painfully slow progress. Having begun with a head start, the NPL researchers found themselves being overtaken by other British teams working at the universities of Cambridge and Manchester. By the time Davies arrived in August 1947, Turing was already terminally disaffected.[2] In the autumn he returned to Cambridge on a year's sabbatical leave from which he never returned, for he was lured from there to Manchester in May 1948 to join the team led by F. C. Williams. Six years later he was dead by his own hand.

Davies, who remained at NPL for his entire working career, spent the year of Turing's death on a Commonwealth Fellowship at – of all places – MIT. He is a hard man to impress at the best of times, and he seems to have been distinctly underwhelmed by MIT. He observed that, while most Institute researchers were passionately interested in devising operating systems for computers, few of them seemed to be concerned about how people might actually use the machines. 'At MIT,' he recalled, 'there was much more interest in the technology of how you made a computer available than in actually using it.'[3]

On his return to Teddington Davies worked on various assignments, including a classified communications project which involved sending data securely over telex links from a weapons-testing range to a processing centre. Then he was drawn into a series of largely futile efforts made by successive British governments to foster an indigenous computer industry which could compete with the Americans. In the 1960s, for example, the then Labour administration launched the Advanced Computer Technology project and, as the senior NPL figure in the field, Davies was nominated to head it. Like most of its predecessors (including ACE), the project eventually fizzled out – despite yielding spin-offs which later became commercial products[4] – but it at least had the effect of ensuring that Davies kept up with developments in the United States.

In May 1965, he went on an extended transatlantic visit. The ostensible excuse was to attend the International Federation for Information Processing (IFIP) Congress in California, but because there was always pressure in the NPL to justify the expenditure of a transatlantic trip, Davies set up a number of visits to various research centres. He went, for instance, to Dartmouth, New Hampshire to look at the development of the BASIC computer language by John Kemeny and Thomas Kurtz – work which impressed him because it was geared towards empowering ordinary computer users. He also visited sites where time-sharing was being done – not because he was particularly interested in the subject, but because it was becoming a hot topic in computer science and Davies was expected to keep abreast of it. Like everyone else, he made the pilgrimage to Project MAC at MIT; less conventionally, he also went to RAND in Santa Monica, where he learned a great deal about the Corporation's JOSS time-sharing project – but (ironically) absolutely nothing about Paul Baran or his design for a survivable network.

Back home at NPL, Davies organised a three-day seminar in November on the subject of time-sharing. The attendees included (somewhat to his surprise) several technical people from the British Post Office research centre at Dollis Hill. Their presence was significant because at that time the Post Office held the monopoly on long-distance telephone communications. Following the NPL meeting there was a one-day open meeting organised by the British

Computer Society and attended by a sizeable contingent from MIT including Jack Dennis, Fernando Corbato, Richard Mills – and Larry Roberts.

These two discussions – the seminar at NPL and the BCS meeting – were what finally concentrated Davies's mind on the problems of data communications involving computers. Time-sharing required having lots of terminals and devices (printers, file-stores and the rest) communicating with a central mainframe computer. Some of these were locally based, and effectively hard-wired to the host machine. But in order to derive the maximum benefit from time-sharing, some of the terminals had to be geographically remote and capable of communicating with the host via the public switched telephone system. And therein, mused Davies, lay the really interesting challenge.

The problem was that there was a mismatch between the way the telephone system worked and the way computers and their ancillary devices communicated. The telephone network was essentially a set of switches for creating connections between one telephone and another. In the early days, the nodes of the network were actually called 'switchboards' because operators manually plugged wires into sockets to set up physical connections between subscribers. Later the operators were replaced by banks of electro-mechanical switches in buildings known as telephone 'exchanges', and later by fully electronic ones.

The telephone network was a *circuit-switched* system. If I wanted to call my mother in Ireland, I first dialled into my local exchange, which then set up a link to a regional exchange which then set up a link to an international switching centre, which then set up a link to its counterpart in Ireland, which then . . . well, you get the idea. Eventually, Ma's phone would ring and when she picked it up it was as if a continuous length of copper cable stretched from my handset in Cambridge to hers in County Mayo. But that was an illusion created by circuit-switching.

Circuit switching was fine for voice communication. But it had several drawbacks. First, it was inherently slow – think of all those switches which had to be thrown in sequence to make the connection. As time and technology moved on, and the switches changed from being clumsy electro-mechanical devices to all-electronic ones, the time needed to set up a call reduced. But still it was slow compared to the operating speeds of computers – even in 1965.

Secondly, circuit-switching was intrinsically uneconomic. Telephone plant – lines, exchanges, switches and so on – costs money to install and maintain. It's an asset which ought to be worked as hard as possible. But while Ma and I were talking, nobody else could use the line. And the circuit had to be maintained for the duration of the call, even if there happened to be long periods of silence during the conversation. This might be appropriate for human conversation – where silences are often part of the interaction, as they were with Ma and me. But for data communications it was idiotic because transactions between computers are intrinsically 'bursty' – they take place in intense, short spurts punctuated by much longer periods of inactivity.

Consider, for example, the communications requirements implied by time-shared computing. The remote user sits at a terminal connected, via a circuit-switched telephone line, to a host computer. For much of the time he – and the circuit (which, remember, is tied up by the connection) – is inactive. The user is thinking, perhaps, or hesitantly typing in best two-finger style. Then he hits the Enter key to send the line of text and there is a frantic burst of data down the (telephone) line, followed again by absolute silence. The host machine absorbs what the user has typed, and responds with another burst of characters which appear on his screen. Then the line goes quiet again. So most of the time that expensive telephone connection is doing precisely nothing. In fact, it was reckoned[5] that in a typical interactive computer session *less than 1 per cent* of the available transmission capacity was being used.

The obvious solution, of course, would be to open the connection just before one device is ready to send a burst, and close it immediately afterwards. But given the rigmarole – and the time delay – involved in setting up a circuit, that was never a feasible proposition, even with the newer generation of electronic switches which began to appear in the 1960s.

There had to be a better way, and one evening in November 1965 it occurred to Donald Davies. What he was seeking was a network tailored to the special needs of data communication. As it happened, data communications networks of a sort already existed. They were called message-switching systems and they were based on an idea as old as the hills – or at any rate as the electric telegraph.

As telegraph technology spread over the globe in the nineteenth century, a method evolved of getting messages from A to Z even if the two points were not serviced by a common telegraph line. What would happen is that telegraph operators at intermediate points would receive a message on one line and retransmit it on another. When a telegraph office had several lines emanating from it (like a hub), it was in effect performing a switching function.[6]

Eventually, teletypes with paper tape punches and readers were developed, and this process of relaying or switching telegrams from one telegraph line to another became semi-automatic. The new machines made it possible for an incoming message to be punched automatically for retransmission on the appropriate outgoing line. When the incoming message had concluded, the operator would tear off the paper tape and then feed it into another teletype. It was known as the 'torn-tape' system.

In 1963, a computerised version of torn tape was introduced when an Iowa company installed a computer-based switch for the airline companies of North America.[7] The system replaced paper tape by storing incoming messages in a disk file and forwarding them automatically to the appropriate outgoing line when it became available. It was the first operational automated 'store-and-forward' message-switching system. Each message had a header containing an address and possibly also routing information so that the message processor at each node could determine which output line to retransmit on. The processor maintained message queues at each outgoing link which were normally serviced on a first-come, first-served basis.

Although simple in principle, message-switching systems were very different from circuit-switched ones. For one thing, the message source and destination did not have to interact in real time. (Think of it in terms of the contrast between sending someone an e-mail and calling them up on the phone.) More importantly, the two types of system behave very differently under stress. An overloaded circuit-switched system will simply refuse to make a connection (you get an engaged or unobtainable tone which prevents you from even dialling). It either serves you or it doesn't. A message-switching system, in contrast, will continue to accept messages even if it is 'busy', but the time to deliver them will begin to lengthen as the load on the network increases. The system degrades, in other words, under stress. But the degradation is not uniform. For a while, message transit time increases more or less in proportion to the increase in traffic. Then, at a certain critical load level, transit time suddenly increases dramatically and the system effectively locks up.

The other problem with conventional message-switching systems was that their performance could be uneven because of the way they handled messages in one piece. Short messages could be infuriatingly held up while the system processed longer ones which happened to be further up the queue. And since each message had to be treated as a whole and had to have its integrity maintained, even if it had passed through a dozen intermediate stations on the way, the nodes had to do a lot of error checking and looking up addresses and storing and retrieving and a hundred other tasks. All of which meant that message-switching systems were totally unsuitable for interactive computing, where speed and reliability of response was essential to sustain the illusion that each user had the machine to him or herself.

Davies's great idea was to break the tyranny of message diversity by carving all messages into small, uniformly sized pieces which he called 'packets'.[8] His initial idea, outlined in an eight-page proposal dated 15 November 1965, seems to have been that each packet would be 400 bits (fifty characters) long,[9] but at some later point it was decided that packets should consist of exactly 1,024 bits (128 characters). These 1,024 bits would contain not just *data* (that is, part of the content of the message) but also a *header* giving information about the packet's source (that is, originator), its destination, a *check digit* enabling the recipient to detect whether the packet had become corrupted in transmission and a *sequence number* indicating where it belonged in the sequence of packets into which the original message had been fragmented.

It was a beautifully simple idea with far-reaching implications. It implied a network with permanent connections through which packets effectively found their own way. Although each node or switch in the network would still perform store-and-forward operations, it would be doing them with uniformly small units of data. All a node needed to be able to do was to read the packet header, parse the destination address and choose an appropriate channel for onward transmission to another node. 'One of my key observations', wrote Davies, 'was that, provided the capacity of the network was adequate, any delay can be kept short by using small packets and high line capacity.' He estimated that with a packet size of 1,024 bits and lines with a capacity of 1.5 megabits/second (the limit of established technology at the time) delays would be 'tens of milliseconds at most'.[10]

At any point in time, each communications line in the network would be carrying packets from many different messages and many different senders, each on its own way through the maze. Packets might take different routes through the network; the route each took was determined by each node's understanding of the state of the network at the moment of forwarding. Similarly, packets might arrive at their destination in a different order from that in which they had been dispatched. But the information encoded in their headers would enable the recipient to reassemble the message in the right order. As each packet arrived, an acknowledgement would be dispatched back to the sender. If no acknowledgement was received, the sender would simply retransmit the packet. And because of the short delay times, the whole crazy thing would work. Not only that, it would be robust, efficient and fast. Davies was confident of that because he had done some simulations, and he was not someone given to wishful thinking.

For the most part, he kept his ideas to himself and a small circle of staff within NPL. But in March 1966 he gave a public lecture in London on 'The Future Digital Communication Network' in which he described his ideas about packet-switching. He was approached afterwards by a man from the Ministry of Defence,

Arthur Llewellyn, who told him about some remarkably similar research by a chap called Paul Baran which had been circulating in the US defence community. It was the first Davies had heard of either the man or his work.[11] When the two met several years later, Davies is reputed to have said: 'Well, you may have got there first, but I got the name.'[12] Henceforth, 'packet-switching' was the name of the game.

By the summer of 1966, Davies had started thinking about building a packet-switched network. In June he wrote a twenty-five-page *Proposal for a Digital Communication Network* in which he set out a design for a system of packet-switching nodes connected by high-bandwidth transmission lines. A packet launched into the system would be of a standard form which included a header stating the addresses of sender and destination. Once accepted by the network, the packet would then be routed to its recipient via intermediate nodes.[13] Davies's proposed system had a certain level of redundancy built into its interconnections, and in that sense resembled Baran's. But he also made provision for something neither Baran nor ARPA had at that time envisaged – a facility for connecting multiple users to the network via intermediate (or 'interface') computers. And he included some detailed performance calculations suggesting that each node ought to be able to process 2,500 packets a second together with an estimate of what each node would cost (£50,000). All of which was just a prelude to proposing that an experimental network with a projected life of five years should be built.

Although Davies's design clearly envisages a *national* network, there was simply no way such a proposal could have been progressed through, let alone implemented by, the government body which at that time held the national monopoly on telecommunications – the General Post Office. Although many of the GPO's technical and research staff were aware of, and enthusiastic about, the ideas coming out of NPL, in the higher reaches of the organisation's bureaucracy the usual blend of inertia, ignorance and indifference reigned. If they were going to press for a UK-wide experiment, therefore, Davies and his colleagues were likely to encounter difficulties similar to those which had caused Baran to abort his proposal a year earlier.

And then Fate intervened. In August, Davies was promoted to head up NPL's Computing Division, a move which gave him the authority needed to make things happen. If he could not have a national network, he reasoned, he could at least build a working model in his own bailiwick. Accordingly, he moved to construct a packet-switched network on the extensive NPL site which would 'serve the computing needs of NPL and be a useful advance in computer organisation, and a working prototype for a national network'.[14] By early 1967 Davies's team, led by Roger Scantlebury, had an outline plan for a local area network to link ten computers, ten high-speed peripheral devices, fifty slow peripherals and graph plotters, six computer terminals and forty teletype terminals. By July the design was far enough advanced for Davies to request £200,000 from the Lab's Steering Committee to start buying kit and building the system. In August, four members of the team (Davies, Keith Bartlett, Scantlebury and Peter Wilkinson) wrote a paper[15] describing their embryonic network to be presented at a symposium that the Association for Computing Machinery (ACM) was staging at Gatlinburg in October. In the parsimonious budgetary regime of the NPL, it was inconceivable that all four authors should go to the symposium, so there was some kind of lottery to choose the presenter. The name that came out of the hat was Roger Scantlebury: he was the guy who was to bring the good news from Teddington to Tennessee.

The Gatlinburg symposium opened on 1 October and ran for three days. There were fifteen papers in all, most on topics unrelated to the concerns of ARPA and the NPL crowd. Larry Roberts's paper[16] was the first public presentation of the ARPANET concept as conceived with the aid of Wesley Clark – that is to say, incorporating the notion of a subnetwork of standardised IMPs which acted as go-betweens for the various host computers attached to the network.

Looking at it now, Roberts's paper seems extraordinarily, well, *vague*. It is three pages long, has three references and three diagrams. Exactly half of the text is taken up with a summary of the reasons for having a network in the first place – sharing computing loads, remote processing, scientific communication and generally making more efficient use of scarce computing resources. It then sets out the case for Clark's subnet idea before finally sketching out the outline of the proposed network. ARPA researchers, it reports, 'have agreed to accept a single network protocol so that they may all participate in an experimental network'. The protocol, Roberts continued,

> is currently being developed. It will conform to ASCII* conventions as a basic format and include provisions for specifying the origin, destination, routing, block size, and sum check of a message. Messages will be character strings or binary blocks but the communication protocol does not specify the internal form of such blocks. It is expected that these conventions will be distributed in final form during September 1967.[17]

The paper goes on to discuss (inconclusively) the relative merits of using dial-up or leased telephone lines operating at 2,000 bits per second for linking nodes.

And that was it! The inescapable conclusion is that Roberts & Co. had not progressed beyond the idea that some kind of message-switching system would be required. And it is clear that the notion of fixed-length message-blocks or packets of the kind envisaged by Baran and Davies had not yet penetrated the Pentagon. Neither Baran nor Davies is cited and the only reference to message length in the paper comes from the networking experiment Roberts and Thomas Marill had carried out at Bob Taylor's behest in 1965. This had shown, he claimed, that 'the average message length appears to be 20 characters' – in other words, message length is decided by users, not specified by the network.[18]

The NPL paper which Scantlebury presented to the symposium was radically different in form and structure. Looking at it now, one is immediately struck by the difference in presentation values. The ARPA paper was typed on an electric typewriter – probably an

---

*Acronym for American Standard Code for Information Interchange – a standard produced in 1963 to assign numerical codes to alphabetical and other characters.

expensive IBM 'golfball' machine; the NPL document, in contrast, was hacked out by a typist on an old-style manual typewriter. And it is *much* more detailed, running to eleven pages of text and six of diagrams. Baran's 1964 report 'On Distributed Networks' is cited as 'the most relevant previous work'. The paper gives a brief outline of the proposed NPL network, then gives a detailed account of the underlying operating principles and of how the network would appear to the user. It describes the packet-switching idea in detail, explains the anatomy of a 1,024 bit packet and presents the results of calculations of network response time and other performance indicators. And it proposes the use of 1.5 megabit per second communication lines rather than the 2 kilobit per second lines envisaged by Roberts.

It's not entirely clear what happened after Scantlebury's presentation. Hafner and Lyon claim that his lecture was the first time Roberts had heard about the packet-switching stuff, or indeed about Baran.[19] Scantlebury remembers discussions with Roberts after his presentation in which Jack Dennis and Peter Denning (two other contributors to the symposium) supported him in arguing the merits of packet-switching in the form being developed at NPL.[20] Whatever the details, there is general agreement that Roberts came away from Gatlinburg persuaded that packet-switching over faster lines[21] was the technology needed to build his network. And when the ARPA crowd returned to Washington they discovered to their chagrin that Baran's RAND papers had indeed been collecting dust in the office. Their author was rapidly co-opted as a kind of unofficial consultant. The prophet of survivable networks discovered, somewhat to his own astonishment, that he was no longer without honour in his own land.

Thereafter, things moved quickly. By August 1968, ARPA had finished the formal 'request for proposals'[22] to build the IMPs and sent it out to the usual suspects, inviting them to bid for the contract to build the network. The first responses came from IBM and Control Data Corporation, both of whom declined to bid on the ground that the network could never be cost-effective!

In the end, six bids were received. Most of the tenderers opted to use a Honeywell minicomputer – the DDP-516, a machine much in favour for military use because it could be 'ruggedised' (that is, toughened for use in hostile environments). The general expectation was that the contract would go to Raytheon, a major Boston defence contractor, and if ARPA had been a conventional Pentagon agency it probably would have gone that way. But ARPA was a maverick outfit, and the contract to build the network went instead to a small consulting firm in Cambridge, Massachusetts which had become legendary as a bolt-hole for academic talent. Its name was Bolt, Beranek and Newman.

BBN – as it is invariably called – was an extraordinary company. It was one of the original technological spin-offs, set up in the 1950s by two MIT faculty members, Richard Bolt and Leo Beranek, to formalise their extensive consulting practice in applied acoustics. From the outset, BBN welcomed academics and graduate students from both Harvard and MIT. 'If you've ever spent any time at either of those places,' said Robert Kahn, an academic who came to the company from MIT,

> you would know what a unique kind of organisation BBN was. A lot of the students at those places spent time at BBN. It was kind of like a super hyped-up version of the union of the two, except that you didn't have to worry about classes and teaching. You could just focus on research. It was sort of the cognac of the research business, very distilled. The culture at BBN at the time was to do interesting things and move on to the next interesting thing. There was more incentive to come up with interesting ideas and explore them than to try to capitalise on them once they had been developed.[23]

In 1957, Leo Beranek had recruited J. C. R. Licklider to BBN, intrigued not so much by his expertise in acoustics as by his growing interest in interactive computing. Shortly after his arrival, Lick asked the company to buy him a computer, and shortly after that Ken Olsen (founder of Digital Computer Corporation) lent him one of the first models in the PDP range to roll off the production line – which is how a firm of acoustics consultants came to be at the cutting edge of the interactive computing business.

In fact, BBN's unplanned move into computing eventually saved it from disaster in the early 1960s when a big acoustics consultancy job (the Avery Fisher Hall at New York's Lincoln Center) went badly wrong. The growing demand for computing expertise filled the gap left in the firm's core business by the Fisher Hall catastrophe, and its executives were constantly on the lookout for computing-related work. Once again, the MIT connection proved crucial. Two of the key computer people at BBN – Frank Heart and Robert Kahn – were both ex-MIT researchers who knew Larry Roberts and had kept in touch with him. Some time in 1967 Roberts let slip to them that ARPA was contemplating funding a network. The BBN guys went back and decided to gamble $100,000 on preliminary design work in the hope that it would land them the contract.

It turned out to be a terrific bet. By the time the ARPA Request for Proposals eventually arrived in August, BBN was already up to speed on the subject. In thirty frantic days, Heart, Kahn and their colleagues produced a 200-page bid which was by far the most detailed proposal submitted to the Agency. It was, in effect, a blueprint for the Net. The award of the contract was a foregone conclusion. All that remained was to build the thing.

The contract started in January 1969. It was worth just over $1 million and required BBN to build and deliver the first four IMPs. The first one was to be shipped to UCLA by Labor Day. In twelve months the network was to be up and running, with four sites online. As the man said, the impossible we do today; miracles take a little longer.

It was one thing to get the intermediate computers – the IMPs – installed and talking to their 'hosts'. But then what? The researchers working for ARPA realised very quickly that BBN was most concerned with getting bits to flow quickly from one IMP to another but hadn't really thought much beyond that requirement. Accordingly a small group of representatives from the various sites was set up. It was called the Network Working Group.

'We had lots of questions,' recalled one of those present, a young UCLA graduate student called Steve Crocker, '– how IMPs and hosts would be connected, what hosts would say to each other, and what applications would be supported.'

No one had any answers, but the prospects seemed exciting. We found ourselves imagining all kinds of possibilities – interactive graphics, cooperating processes, automatic data base query, electronic mail – but no one knew where to begin. We weren't sure whether there was really room to think hard about these problems; surely someone from the East would be along by and by to bring the word. But we did come to one conclusion: We ought to meet again. Over the next several months, we managed to parlay that idea into a series of exchange meetings at each of our sites, thereby setting the most important precedent in protocol design.[24]

From the outset, the Network Working Group was peculiar. It was largely comprised of graduate students, those vulnerable and insecure apprentices upon whom much of the edifice of scientific research rests. Three founder members of the group – Vint Cerf, Crocker and Jon Postel – had been precocious contemporaries at the same Californian high school. Not surprisingly, they felt they had no formal status in the ARPA project. They guessed they had been assembled to do some of the groundwork, and that later on some really qualified honchos would appear to put it right. They assumed, in other words, that Somewhere Up There was a cohort of guys who really knew what they were doing. In assuming this, they were completely wrong, because the truth was that nobody Up There knew any more than they did. The future of the Net rested in the hands of these kids.

In February 1969 they met with the people from BBN. 'I don't think any of us were prepared for that meeting,' wrote Crocker.

The BBN folks . . . found themselves talking to a crew of graduate students they hadn't anticipated. And we found ourselves talking to people whose first concern was how to get bits to flow quickly and reliably but hadn't – of course – spent any time considering the thirty or forty layers of protocol above the link level. And while BBN didn't take over the protocol design process, we kept expecting that an official protocol design team would announce itself.

A month later, after a meeting in Utah, it became clear to the students that they had better start recording their discussions. 'We had accumulated a few notes on the design of DEL and other matters,' recalled Crocker, 'and we decided to put them together in a set of notes.'

> I remember having great fear that we would offend whomever the official protocol designers were, and I spent a sleepless night composing humble words for our notes. The basic ground rules were that anyone could say anything and that nothing was official. And to emphasize the point, I labeled the notes 'Request for Comments.' I never dreamed these notes would be distributed through the very medium we were discussing in these notes. Talk about Sorcerer's Apprentice![25]

Steve Crocker's first 'Request for Comments' went out by snail mail on 7 April 1969 under the title 'Host Software'. RFC1 described how the most elemental connections between two computers – the 'handshake' – would be achieved.

The name RFC stuck, so that even today the way the Internet discusses technical issues is still via papers modelled on Crocker's RFC idea.[26] It wasn't just the title that endured, however, but the intelligent, friendly, co-operative, consensual attitude implied by it. With his modest, placatory style, Steve Crocker set the tone for the way the Net developed. 'The language of the RFC', wrote Hafner and Lyon, 'was warm and welcoming. The idea was to promote cooperation, not ego. The fact that Crocker kept his ego out of the first RFC set the style and inspired others to follow suit in the hundreds of friendly and cooperative RFCs that followed.'[27]

RFC3, entitled 'Documentation Conventions' and also issued in April 1969, captures the flavour of the enterprise. 'The Network Working Group', wrote Crocker,

> seems to consist of Steve Carr of Utah, Jeff Rulifson and Bill Duvall at SRI, and Steve Crocker and Gerald Deloche at UCLA. Membership is not closed.
>
> The Network Working Group (NWG) is concerned with the HOST software, the strategies for using the network, and initial experiments with the network.
>
> Documentation of the NWG's effort is through notes such as this.

Notes may be produced at any site by anybody and included in this series.

The content of a NWG note, he continued,

> may be any thought, suggestion, etc. related to the HOST software or other aspect of the network. Notes are encouraged to be timely rather than polished. Philosophical positions without examples or other specifics, specific suggestions or implementation techniques without introductory or background explication, and explicit questions without any attempted answers are all acceptable. The minimum length for a NWG note is one sentence.
>
> These standards (or lack of them) are stated explicitly for two reasons. First, there is a tendency to view a written statement as ipso facto authoritative, and we hope to promote the exchange and discussion of considerably less than authoritative ideas. Second, there is a natural hesitancy to publish something unpolished, and we hope to ease this inhibition.

From the outset, it was a requirement that any author of an RFC had to send copies to named researchers at BBN, ARPA, UCLA, Utak and the University of California at Santa Barbara (UCSB).

The NWG's task was to hammer out a set of agreed conventions which would govern exchanges between computers on the network. It was clear that as a matter of urgency the students had to devise a method of enabling network users to log in to a remote machine (the facility later christened 'TELNET') and a facility for securely transferring files from one machine to another across the Net. But they were also acutely conscious of the need to lay a secure foundation for the much wider range of technical standards that would be needed if the embryonic network were ever to achieve its full potential. The trouble was that the computer operating systems of the time were all designed on the principle that each machine was the centre of the universe – communicating with computers made by rival manufacturers was out of the question! And time was pressing – the first IMP was due to be delivered to UCLA on 1 September 1969, and the rest were scheduled at monthly intervals.[28] It was a very tall order.

The TELNET and File Transfer (FTP) protocols which were the first outcomes of the NWG discussions were conceived as *client–server* conventions – that is to say, they envisaged exchanges in which a program running on one computer (the client) requested a service from another computer (the server). To that extent, they were *asymmetrical* protocols. But when Steve Crocker and his colleagues presented their thinking to Larry Roberts of ARPA at a meeting in December he insisted that something more was required. The NWG went back to the drawing board and came back with a symmetric host-to-host protocol which they called the Network Control Program.[29] This was the first ARPANET inter-process communication software and – after the routing software which drove the IMPs – the key program on the Net. At some point later it became known as the Network Control Protocol or NCP.

Why 'protocol'? It seems an unlikely term to have figured in the vocabularies of engineering students, even bright ones, at the time. Vint Cerf, one of the students involved in the NWG, told Peter Salus that it had originated in a conversation between Steve Crocker and Jon Postel, who had thought of it in terms of diplomats exchanging handshakes and information.[30] It was only later that they discovered its etymology (the word comes from the Greek, and means the first page of a document bearing the title, name of the author, scribe and so on).

The RFC archives contain an extraordinary record of thought in action, a riveting chronicle of the application of high intelligence to hard problems, preserved in the aspic of plain ASCII text. Few great engineering projects leave such a clear audit trail. Salus traces the progress of one discussion – the design of the NCP – through twenty-four RFCs spread over twenty-five months. It was this extraordinary co-operation that made the protocols – even the early ones – as good as they became. 'Here we had', wrote Salus,

two dozen people all over the United States putting in their thoughts in a technical free-for-all that would have been unheard of in other quarters . . . Most of the protocols, when they finally appeared as RFCs, bear one or two names at the head. In general, however, each protocol was the result of many good minds at work, in meetings, over the phone, and over the (evergrowing) network.[31]

What these kids were inventing, of course, was not just a new way of working collaboratively, but a new way of creating software. The fundamental ethos of the Net was laid down in the deliberations of the Network Working Group. It was an ethos which assumed that nothing was secret, that problems existed to be solved collaboratively, that solutions emerged iteratively, and that everything which was produced should be in the public domain. It was, in fact, the genesis of what would become known much later as the Open Source movement.

Let us pass over the saga of how the IMPs were built, configured, tested and shipped – not because it's boring (it isn't, and Hafner and Lyon tell the story rather well) but because we can't do everything – and spool forward to the Saturday before Labor Day when the first IMP was shipped to UCLA. It was about the size of a refrigerator, weighed over 900 lb and was enclosed in battleship-grey steel, exactly to military specifications – to the point where it even had four steel eyebolts on top for lifting by crane or helicopter. The UCLA crew manhandled it into its assigned room next to the UCLA host machine – a Scientific Data Systems Sigma-7 – and switched it on. The IMP worked straight out of the crate, but for the moment it had nothing to talk to except the Sigma-7. It takes two nodes to make a network.

A month later, on 1 October 1969, the second IMP was delivered to Stanford Research Institute and hooked up to its SDS 940 time-sharing machine. With both IMPs in place and both hosts running, a two-node network existed – at least in principle. The moment had arrived to see whether it worked in practice.

What happened was the kind of comic event which restores one's faith in the cock-up theory of history. The Los Angeles (UCLA) and Stanford (SRI) sites were linked by telephone, so human dialogue accompanied this first step into the unknown. It was decided that the UCLA side would try to log on to the SRI machine. Years later, Leonard Kleinrock, in whose Los Angeles lab the first IMP had been installed, related what happened in an e-mail message to the *New Yorker* writer John Seabrook:

> As soon as SRI attached to its IMP, under my directions, one of my programmers, Charley Kline, arranged to send the first computer-to-computer message. The setup was simple: he and a programmer at SRI were connected via an ordinary telephone line and they both wore headsets so they could talk to each other as they observed what the network was doing. Charley then proceeded to 'login' to the remote SRI HOST from our UCLA HOST. To do so, he had to literally type in the word 'login'; in fact, the HOSTS were smart enough to know that once he had typed in 'log', then the HOST would 'expand' out the rest of the word and add the letters 'in' to it. So Charley began. He typed an 'l' and over the headset told the SRI programmer he had typed it (Charley actually got an 'echo' of the letter 'l' from the other end and the programmer said 'I got the l'.) Then Charley continued with the 'o', got the echo and a verbal acknowledgement from the programmer that it had been received. Then Charley typed in the 'g' and told him he had now typed the 'g'. At this point the SRI machine crashed!! Some beginning![32]

I love this story. It is about a hinge of history; and yet the drama is undermined by farce which brings everything back to earth. It was the beginning of the wired world – a moment as portentous in its way as when Alexander Graham Bell muttered 'Mr Watson, come here, I want you' into his primitive apparatus. When Charley Kline typed his L he was taking mankind's first hesitant step into Cyberspace. And the first result of his step was a crash!

9:
Where it's @

Letters of thanks, letters from banks,
Letters of joy from girl and boy,
Receipted bills and invitations
To inspect new stock or to visit relations,
And applications for situations,
And timid lovers' declarations,
And gossip, gossip from all the nations.

W. H. Auden, 'Night Mail', 1936

In September 1973, the ARPANET extended an arm across the Atlantic. The organisers of a conference at Sussex University in Brighton had arranged to demonstrate the transmisison of packets via satellite. The link went from Virginia in the US via satellite to Goonhilly in Cornwall, thence via a leased land-line to University College, London and then via another land-line to Brighton, where people could use the ARPANET as if they were sitting in Leonard Kleinrock's lab in Los Angeles.

At the moment which interests us, Kleinrock was in fact sitting in his lab. He had just returned from Brighton (he left the conference a day early) and discovered that he had left his razor behind. History does not record whether it was a special razor, or whether it was just an excuse to try something new, but whatever the reason Kleinrock decided to see if he could get one of the other US delegates to retrieve the thing for him. It was 3 a.m. in Britain, but he found that

Larry Roberts from ARPA was logged in at Brighton. Using a programme called TALK which split the screen and allowed two people to 'converse' in real time by typing, Kleinrock asked Roberts to retrieve his razor. The next day it was ferried to LA by another conference delegate.[1]

Was this the first recorded example of electronic mail? Not really. For one thing, users of time-shared computers had long been passing messages to other users of the same machines. Real electronic mail, in contrast, passes between users of *different* machines. For another, the exchange about Kleinrock's razor was synchronous – the interactions between sender and recipient took place in what engineers call 'real time', whereas the essence of e-mail is that it is *asynchronous*: sender and receiver do not need to be hooked up simultaneously. It was, in fact, an early example of online 'chat'.

E-mail came as something of a surprise to ARPA – or at least to some people in it.[2] The Net was funded, remember, to enable the research community which was funded by the Agency to share scarce resources. The idea was that researchers at one site would be able to log into – and run programs on – computers at other sites. So imagine ARPA's puzzlement when a specially commissioned survey in 1973 revealed that *three-quarters* of all the traffic on the Net was electronic mail. Here was this fantastic infrastructure built at government expense for serious purposes – and these geeks were using it for sending messages to one another![3] Worse still, some of the messages were not even about computing! *Sacré bleu!*

I can well imagine the outrage this must have caused in some quarters. A couple of years earlier I had been a research student at Cambridge. Like most of my peers, I was a dedicated user of the university's time-sharing system. I saw it as a gigantic calculator – a machine whose sole purpose was to crunch numbers – in my case to solve the differential equations which made up the simulation model I was building.

Because the Control Lab where I worked did not have its own printer, I had to bicycle to the Computer Lab on the Old Cavendish site to collect the output from the simulation runs of my model. The wodges of printout were periodically taken from one of the line-printers and placed in large folders – called 'tanks' for some reason – arranged in alphabetical order. I remember one day picking up my twenty or so pages and noticing that the tank next to mine was bulging with an enormous mass of printout. Curious to see what kind of number-crunching could produce such prodigious output I sneaked a look – and found to my astonishment that the pages contained not numbers but *words* – that in fact what I was looking at was a draft of a PhD dissertation.

I was shocked and outraged by such a frivolous use of computing resources. Here was this stupendously expensive and powerful mach-ine,[4] and some bloody research student was using it *as a typewriter!*

Viewed from today, when probably 95 per cent of all computer time is spent word-processing, my youthful indignation seems quaint. It was however less incomprehensible at the time. Remem-ber that we had been heavily conditioned by our experiences of computing. We stood in awe of the technology. Computers were very expensive devices. Even universities like Cambridge could only buy them by special leave of a powerful committee of the University Grants Committee called the Computer Board. Computers required air-conditioned rooms and were surrounded by an elaborate priest-hood. For a mere graduate student to be allowed to use such a resource seemed an incredible privilege – not a right. And the privilege carried with it an obligation to employ the machine for serious purposes. Using it as a typewriter definitely did not qualify.

I would have felt much the same about electronic mail – had I known about it. In fact I discovered e-mail about two years later – about the time Kleinrock was retrieving his razor. By that stage, all time-shared systems had mail as one of their facilities – though of course it was purely internal because it extended only to users of the system. I discovered it first when the Help-desk telephone was permanently engaged and someone on the next console suggested sending the duty adviser a message. But even then I approached it with due seriousness. The idea that one might send an e-mail message conveying an endearment or the result of a football match or a joke – or that a Hollywood studio would one day release a major feature film about two characters who fall in love through e-mail[5] – would have seemed sacrilegious.

*

And now? The first thing I do when I get down in the morning to make a cup of tea is to switch on my computer and set it to collect my e-mail. It's often the last thing I do at night, before I switch the machine off and climb the stairs to bed. In between I log in ten or twelve times to see what's in my in-box. And what's there is an amazing variety of things. I get a digest of several daily newspapers, for example. I get messages from magazines with the contents list of the latest issue. Every day there are newsletters from online publishers and news outlets containing what their editors regard as the day's hot stories about developments in the computing industry.

There are messages from readers of my newspaper column contesting my views, pointing out errors or asking for more information. There are e-mails from work colleagues commenting on drafts, arranging meetings, asking for information, gossiping and grousing and generally chewing the organisational cud. And there are notes from a wide circle of e-mail correspondents – people I know slightly or well in real life, but with whom I have intensive, electronically mediated relationships, in some cases extending over years: an academic colleague in Holland; a neighbour who lives 100 yards away but always seems to be away when I am around; a former research collaborator who now works in the US; an American friend who makes a rich living as a corporate raider; a distinguished American academic lawyer; and a few journalists who understand the other side of my professional existence.

It's almost impossible to explain the hypnotic attraction of e-mail to non-wired folks, so I've largely given up trying. It's no use telling them that I've been using it from home since 1975 (because they never believe me). Nor is it any good explaining that for me it's a much more important medium than the telephone (because they roll their eyes and check the line to the nearest exit). So now I just shrug when people ask me about it. And when they eventually take the plunge and come back raving to me about the wonders of e-mail, I also shrug. There's no percentage in saying 'I told you so.' It's one of those things where experiencing really is believing.

What makes e-mail special is that it's a strange blend of writing and talking. Although messages are typed, most of them read like stream-of-consciousness narratives, the product of people typing as fast as they can think. They are often full of typos and misspellings, and niceties like apostrophes often fall by the wayside. Indeed I sometimes think that the missing apostrophes are the key to understanding it. Sitting reading e-mail messages I am often reminded of the letters and diaries of Virginia Woolf, surely one of the greatest correspondents and diarists of the twentieth century. Her private jottings have the same racing immediacy, the same cavalier way with apostrophes, the same urgency. Here she is, for example on Saturday, 28 December 1929:

> Bernard Shaw said the other night at the Keynes – & the Keynes's have just wrecked my perfect fortnight of silence, have been over in their rolls Royce – &L. made them stay, & is a little inclined to think me absurd for not wishing it – but then Clive is his bugbear at present – Bernard Shaw said to me, I have never written anything but poetry. A man has written a book showing how by altering a word or two in a whole act of the D[octo]rs Dilemma is in rhythm. In fact my rhythm is so strong that when I had to copy a page of Wells the other day in the midst of my own writing my pen couldnt do it.; I was wanting to write my own rhythm – yet I hadn't known till then that I had so strong a rhythm. Heartbreak House is the best of my plays. I wrote it after staying with you at the Webbs in Sussex – perhaps you inspired it. And Lydia is making the Lady into Queen Victoria. No I never saw Stevenson – Mrs Stevenson thought I had a cold.[6]

And here is an e-mail from a friend of mine:

From: G

To: John Naughton
Subject: RE: Parcels
Date: 08 July 1996 02:43

Dear John,
I was v. pleased to see yr piece about the civil risks of Europomp, esp. as I had a nightmare experience this weekend. No danger, but horrors of many kinds – all lurking at, where else, but the Three Tenors concert. Classic FM invited me; I took J.D. as my guest; our

tickets said L350 each; the crowd was white, (except for the people from premier sponsors JAL) middle class, (lower and upper) all jam packed into these minute plastic seats. It was all so solemn, so gigantic, so remote from anything to do with singing. People came stamping in an hour after the due start time, (it sounded like drum beats at onepoint.) Phones rang. Everyone smoked. It began to rain. D and I left – hastily – although we both had sensible mackintoshes. We agreed we had just met Thatcher's people en masse and we would be unlikely to meet any of them again at any other event. Horrid. Like a very bad dream. House full of young male persons yesterday, some going on hols, some coming back. Very cheering. I found Primary Colours rather a slog. I suspect all that detail means it was a wumman wot writ it. Love, G

Note the time – 2.43 a.m.: this is a message dashed off at the end of the evening in question, while its outrages, quirks and pleasures were still fresh in the mind.

These two passages are separated by sixty-seven years and a world of social change, yet they clearly belong to the same genre – a highly personalised, subjective, compressed kind of reportage which blends external observations with private experience and eschews the typographic conventions usually employed to distinguish between one and the other.

Woolf would have loved e-mail, and not just because it provides the immediacy and intimacy her style craved, but also because it makes possible a degree of epistolary interaction not seen since the days of the Edwardian postal system. I remember once reading somewhere an account of her (in London) exchanging three – or was it four? – letters with E.M. Forster in Cambridge in a single day. Nowadays, such a performance would be difficult or impossible. But with e-mail it's easy.

The greatest thing about electronic mail is the fact that you can reply instantly. If you receive a piece of physical mail – a letter or a postcard – sending a reply always involves a certain amount of bother.[7] But with e-mail you can do it on impulse and then hit a button and it's off. 'You do not have to print out a piece of paper,' writes one seasoned exponent of the art, Robert M. Young, 'put it in an envelope, put a stamp on it, take it out of the house to a pillar box and put it in.'

> It occurs without that whole string of moments during which we can potentially think again. You don't have to ask, 'Is this a message I am prepared to have in print in someone else's house?' The facts that they can print it out, that it can be archived on their hard disc or even in the records of an email forum and are thereby accessible to total strangers – even more so than a private letter would be – somehow do not come to mind. Once again, not having to take an envelope outside the house is important. It eliminates (or at least mitigates) the dynamics of separation anxiety, of objectification, of a real-world action. It is almost as if it is all occurring in the head. There is no physical artefact outside the screen.[8]

The strangest thing about electronic mail is that a medium which is entirely enabled by machines should feel so intimate. 'I have had total strangers write to me about the most intimate matters,' says Young.

> I have had them say very flattering or very critical things of a sort I have rarely received in letters. I have had highly erotic (though rarely sexual in the narrow sense) letters from men and women. I have received extraordinarily intemperate letters (and have sent one or two) from strangers, from acquaintances, from friends, from colleagues. It is [as] if they don't 'count' in the same way letters on letterhead, or even phone conversations, do.[9]

The ease with which one can reply to messages – and copy or forward them to others – can of course sometimes lead to terrible *faux pas*. The problem is that it's too easy. Every e-mail junkie I know has experienced at least once that terrible sinking feeling you get when you realise you've just copied a particularly frank appraisal of your editor/boss/lover/subordinate to the person in question. Or to the entire organisation in which you work.

A couple of years ago an online news agency ran a feature in which subscribers confided some of the disasters which can happen with e-mail. One – a magazine editor – related how an important advertiser bombarded her with messages requesting a meeting to talk about their product. 'Feeling a little punchy and not wanting to go to the meeting,' she recalled, 'I sent out the letter to one of my co-workers with the comment, "I'd rather suck raw eggs" (or something slightly ruder) – and then clicked on the "Reply" rather than the "Forward" button. You can imagine the sequel.'

Another respondent related how, as a joke, he made a recording of a friend snoring, stored it as a sound file on his computer and was about to send it as an e-mail attachment to a colleague. (People do the strangest things.) Startled by someone entering his office, he hastily clicked 'Send' – and then discovered that he had actually clicked 'Send to All', thereby distributing the message to everyone on the network. 'As I frantically tried to stop the message from going out,' he reported,

> I heard a person laughing in another cubicle, and then a few more, and then almost everybody. I was so embarrassed that I dropped my coffee on my desk, which shorted out my computer. About the same time, my friend opened the email and realised that he's the person snoring in the sound file. The next thing I know, I'm in the boss's office, getting a decrease in salary for two months.

There's lots more in the same vein. E-mail is wonderful – but think before you click. Otherwise you may wind up repenting at leisure.

'Real' e-mail dates from 1970.[10] There had been 'internal' electronic mail on time-shared systems for almost a decade before that: MAILBOX, for example, was installed on MIT's Project MAC system in the early 1960s. But the first machine-to-machine mail transfer took place in July 1970, between two Digital computers in Bolt, Beranek and Newman's lab in Cambridge, Massachusetts.[11] The guy who made it work – and the man who deserves to be remembered as the inventor of electronic mail – was a BBN hacker called Ray Tomlinson.

Tomlinson, needless to say, was an MIT man. After graduating from the Institute in 1965, he spent two years studying for a doctorate, then went to BBN. The computers he was working on were PDP-10s, minicomputers which ran a time-sharing operating system and therefore supported multiple users. Tomlinson had written a mail program which enabled one user to send mail to another user of the same machine. To send, you ran a program called SNDMSG; to receive mail you ran a program called READ-MAIL. Looking at the two machines sitting in the lab, Tomlinson decided to see what was needed to send mail between them. He did so by modifying an experimental program called CPYNET which he had written some weeks earlier for transferring files between the machines. The modified CPYNET was designed to carry mail messages, and when Tomlinson tried it, it worked.

Unlike Alexander Graham Bell's famous first call to his assistant, Watson, the content of the first e-mail message – from Tomlinson on one computer to himself on another – is forgotten. What we do know is that Tomlinson didn't make a big deal of his breakthrough. 'When he showed it to me,' one of his colleagues, Jerry Burchfiel, recalled, 'he said, "Don't tell anyone! This isn't what we're supposed to be working on." '[12]

It was one thing to get mail running between two identical machines in the same lab; but how to make it work between the disparate machines on the Net? The solution was to adapt the method Tomlinson had used. In 1972, the ARPANET community was finalising FTP – the File Transfer Protocol to be used for transferring files around the Net. Someone suggested piggy-backing Tomlinson's mail program on to the end product. By the autumn of 1973 a mail protocol was effectively agreed[13] and his genie was out of the bottle.

Tomlinson is also responsible for creating what later became one of the icons of the wired world – the use of the symbol @ as a device for separating the name of the sender of an e-mail message from the network-ID of the machine on which s/he had a mailbox. He got it from examining the keyboard of the Model 33 teletype he was using at the time, looking for a punctuation mark which could not conceivably form part of anyone's name, and his gaze alighted on the @ key. It seemed an appropriate choice because it signified 'at', but it gave rise to unanticipated problems because in the MULTICS operating system used at MIT it just so happened that @ was the character used to send a 'line kill' command. In other words, if you were a MULTICS user and found that halfway through a line you had made a mistake, pressing the @ key would helpfully delete the entire line. This of course also meant that MULTICS users who tried to enter e-mail addresses of the form somebody@bbn-tenex would find that the operating system kept throwing them away before they even got to type the message!

The @ problem was a side-show, however, compared to the 'header wars' which raged for several years in the closed world of ARPANET users. Every e-mail message needs a header indicating the sender's identity and address, the date and time of dispatch, the name and address of the addressee etc. Given that a typical message may pass through numerous different e-mail gateways as it wings its way to the destination, there needs to be consistency in the format and syntax of these headers. But there were as many different views about what headers should contain as there were people sending messages. Some people argued that headers should include data like keywords, character and word counts, and so on. Others argued for economy and terseness on the ground that most e-mail messages were short and would be unbalanced by elaborate headers. The problem was resolved in 1975 when Ted Myer and Austin Henderson from BBN issued RFC 680, 'Message Transmission Protocol', a new prescription for standard headers around which the Net could regroup. The protocol was revised again towards the end of 1977 and has remained substantially unchanged ever since.

The final – and perhaps the most important – breakthrough in the design of e-mail systems came from a programmer called John Vittal, who had written a popular program called MSG in 1975. His great contribution was to include an 'answer' command in his e-mail reader. From then on, replying to an incoming message was as simple as hitting a key. Electronic mail as we know it had arrived.

E-mail is still the dominant use of the Net in terms of what most people do with it every day, though it no longer accounts for the bulk of the data traffic. Since 1993 e-mail has been rather over-shadowed by the World Wide Web (see later) and the associated hype about multimedia, and now seems low-tech, humdrum, ordinary – almost not worth bothering with.

Of these, the only justifiable adjective is low-tech: e-mail has minimal technical requirements; you need a computer, of course –

but almost anything will do, even an elderly DOS-box – and a modem, some software and an account with an Internet Service Provider which gives you an electronic mailbox.

And, by and large, e-mail is very light on telephone bills. There are terrific software packages like Eudora and Pegasus Mail[14] which enable you to compose and read your mail offline. When you've finished writing, you simply hit a button, the program dials the ISP computer, uploads all your queued messages in a quick burst, downloads any new messages in your mailbox, and hangs up. With such a program you could do an awful lot of e-mail correspondence in thirty minutes of connect time *over an entire year*.

The Net was built on electronic mail. To ARPA's surprise, it proved to be the prime mover in the network's early growth and development. 'E-mail was to ARPANET', wrote Hafner and Lyon, 'what the Louisiana Purchase was to the young United States.'[15] Likewise the metamorphosis of the ARPANET into the Internet, with its proliferating structure of global conferencing and news groups, was all based on e-mail. It's the oil which lubricates the system. And the great thing is that because it's based on the apparently inexhaustible desire of human beings to communicate with one another, it's an infinitely renewable resource.

10:
Casting the Net

Network. Anything reticulated or decussated at equal
distances, with interstices between the intersections.
> Samuel Johnson, *A Dictionary of the English Language,* 1755.

T he strange thing about the ARPANET was that it worked
more or less as advertised from the word go. Bolt,
Beranek and Newman delivered IMPs at the rate of one a month
and the network grew apace. The University of California at Santa
Barbara got the third one in November 1969 and Utah took delivery
of the fourth early in December. IMP number 5 was delivered to
BBN itself early in 1970 and the first cross-country circuit – a 50
kilobit per second line from Leo Kleinrock's machine in Los Angeles
to BBN's in Boston – was established. This meant not only that the
Net now spanned the continent, but also that BBN was able to
monitor it remotely.

By the summer of 1970, the graduate students on the Network
Working Group had worked out a provisional version of the
Network Control Program (NCP) – the protocol which enabled
basic communications between host computers – and IMPs 6, 7, 8
and 9 had been installed at (respectively) MIT, RAND, System
Development Corporation and Harvard. Towards the end of the
summer, AT&T (whose engineers were presumably still baffled by
the strange uses the researchers had discovered for telephone lines)
replaced the UCLA–BBN link with a new one between BBN and

RAND. A second cross-country link connected the University of Utah and MIT. By the end of 1971 the system consisted of fifteen nodes (linking twenty-three hosts). In August 1972 a third cross-country line was added.[1] By the end of the year ARPANET had thirty-seven nodes. The system was beginning to spread its wings – or, if you were of a suspicious turn of mind, its tentacles.

Researchers were also beginning to get a realistic idea of what you could do with it. They could log in to remote machines, for example, and exchange files securely.[2] Later in 1970 they also saw the first stirrings of electronic mail. And, as always, there were some restless kids who wanted to do some wacky things with their new toy. In January 1971, for example, two Harvard students, Bob Metcalfe and Danny Cohen, used a PDP-10 minicomputer at Harvard to simulate an aeroplane landing on the flight deck of an aircraft carrier and then passed the images to a graphics terminal located down the Charles River in MIT. The graphics were processed on the MIT machine and the results (in this case the view of the carrier's flight deck) were then shipped back over the Net to the PDP-10 at Harvard, which displayed them. It was the kind of stunt which makes non-technical people wonder what kind of stuff students are on, but in fact the experiment had a serious purpose because it showed that the Net could move significant amounts of data around (graphics files tend to be large) at a rate which approximated to what engineers call 'real time'. Metcalfe and Cohen wrote an RFC[3] describing their achievement under the modest header 'Historic Moments in Networking'.

The emerging ARPENET was a relatively closed and homogeneous system. Access to it was confined to a small elite working in Pentagon-funded computing laboratories. And although it wove together a number of varied and incompatible mainframe computers (the hosts), the subnetwork of IMPs which actually ran the network was comprised of identical units controlled, updated and debugged from a single Network Control Centre located in Bolt, Beranek and Newman's offices in Boston. The subnetwork showed its military colours in that it was designed to be exceedingly reliable in its performance and behaviour. Indeed, one of the reasons the designers chose the Honeywell 516 as the basis for the IMPs was

that it was a machine capable of being toughened for military use –
which meant that it could also (in theory) be secured against the
curiosity of graduate students. (This was before the folks at ARPA or
BBN realised how crucial those students would be in getting the
damn thing to work.)

The modern Internet is significantly different from the system
that BBN built. In the first place, it is comprised of an unimaginable
variety of machines of all ages, makes and sizes, running a plethora
of operating systems and communications software. Secondly, it is
in fact *a network of networks*: the components of the modern
Internet are themselves wide-area networks of various sorts. There
is no Network Control Centre probing the nodes, installing software updates from afar and generally keeping an eye on things. And
yet, in the midst of all this headless, chaotic variety, there is order.
The system works. Packets get from one end of the world to the
other with astonishing speed and reliability.

If you have any doubts, try this. I have on my computer a lovely
little program called PING. Its purpose is to send out test packets to
a destination anywhere on the Net in order to test how reliably they
reach their destination and how long they take in transit. Now let's
PING a node in San Francisco: it's www.kpix.com, the site which
provides those live pictures of the Bay Area which first captured my
attention. After twenty-two pings, I call a halt and the program
summarises the results in a table:

www.kpix.com	204.31.82.101	
Sent=22	Received=20	Packet loss=9.09%
Min rtt=441	Max rtt=650	Avg rtt=522

The first row indicates that the site address has been translated into
the underlying Internet addresses – the set of four numbers which
uniquely identifies the computer to be PINGed. The second row
shows that twenty of the twenty-two packets (that is, 90.91 per
cent) reached their destination.[4] The third row reveals that the
minimum time for the round trip was 441 milliseconds (thou-
sandths of a second), the maximum was 650 and the average
worked out at 522. That is to say, it took, on average, just over half

a second for a packet to travel from my study in Cambridge, UK to KPIX's machine in San Francisco and back again.

Looking back at the ARPANET from the vantage point of the contemporary Net is a bit like comparing a baby chimpanzee with a human child. The resemblances are striking and the evolutionary link is obvious. It is said that something like 98 per cent of the genetic material of both species is indistinguishable. Yet we are very different from chimps.

So it is with the two networks. The Internet has inherited many of its dominant characteristics from its pioneering ancestor, but it is significantly different in one respect: its ability to diversify. ARPA-NET could never have expanded the way the Internet has: its design required too much control, too much standardisation for it to match the variegated diversity of the online world. For it to metamorphose into the network which now girdles the globe, something had to change.

In a way, the ARPANET was a 'proof of concept system'. It took a set of ideas many people thought impracticable, and created a working network out of them. By 1972 it was clear that the project was, in technical terms, a runaway success. But, as many inventors know, designing a better mousetrap is no guarantee that the world will beat a path to your door. Even geniuses have to blow their own trumpets sometimes.

Within BBN, one man – Robert Kahn – understood this better than most. He was born in Brooklyn, New York in 1934, got a Bachelor's degree in electrical engineering from City College of New York and then went to Princeton, where he picked up a Master's and a doctorate. In 1964 Kahn went to MIT as an assistant professor, and two years later took temporary leave of absence (as many MIT engineering faculty members did) to work at BBN. He never made it back to MIT, because when the company decided to bid for the ARPANET contract Kahn was persuaded to take on the job of overall system design and was hooked for life.

Kahn was a systems engineer, not a hacker. While others were concerned with the individual components which made up the network, he was the guy who knew that systems were more (and often less) than the sum of their parts – that the whole network would behave in ways that could not be predicted from a study of its individual components. His professional instincts told him, for example, that the routing algorithms – the procedures which governed the way the IMPs processed packets – would be critical. 'It was my contention', he recalled, 'that we had to worry about congestion and deadlocks.'

> What do you do when the network just fills up? Things might come to a grinding halt. I was busy at work designing mechanisms that would prevent that from happening or, if it did happen, that would get you out of it. And the prevailing feeling of my colleagues was it's like molecules in a room; don't worry about the fact that you won't be able to breathe because the molecules will end up in a corner. Somehow there will be enough statistical randomness that packets will just keep flowing. It won't be able to somehow block itself up.[5]

Determined to prove the optimists wrong, Kahn and a colleague named Dave Walden went out to California early in 1970 to test the fledgling Net to destruction. 'The very first thing we did', he remembers, 'was run deadlock tests. And the network locked up in twelve packets. I had devised these tests to prove that the network could deadlock. There was no way to convince anybody else, particularly the person writing the software, that the network was going to deadlock – except by doing it.'[6] From the outset Kahn had been of the view that the network project should be a large-scale experiment, not something cooked up in the rarefied atmosphere of an individual lab or a number of geographically proximate institutes. He felt that limited systems might not scale up and wanted a continent-wide network, using long-distance phone lines, from the word go. And his view prevailed, which is why from early in the project the ARPANET spanned the continental United States.

In the middle of 1971 Kahn turned his mind to the problem of communicating this astonishing technical achievement to the movers and shakers of the US political, military, business and telecommunications communities. After a meeting at MIT of some of the main researchers involved on the project, it was decided to organise a large-scale, high-profile, live demonstration of what the network could do. Casting round for a suitable venue, Kahn hit on the first International Conference on Computer Communication, scheduled to be held in the Washington Hilton in October 1972, and negotiated with the organisers an agreement that ARPA could mount a large-scale live demonstration of the network in action.

The goal of the exhibition was to be the most persuasive demo ever staged – 'to force', in Kahn's words, 'the utility of the network to occur to the end users'.[7] It was to be the event that made the world take notice of packet-switching, because up to that point the technology had been more or less invisible outside the elite circle of ARPA-funded labs. 'A lot of people were sceptical in the early days,' said Kahn. 'I mean, breaking messages into packets, reassembling them at the end, relying on a mysterious set of algorithms, routing algorithms, to deliver packets. I'm sure there were people who distrusted airplanes in the early days. "How are you going to ensure that they are going to stay up?" Perhaps this was the same kind of thing.'[8]

Kahn and his team put an enormous effort into mounting the demo. They toured the country bullying, cajoling, tempting researchers and computer equipment manufacturers into participating. On the days before the conference opened, hackers gathered from all over the country to begin the nightmarish task of assembling and testing all the kit in the designated hall of the hotel. The atmosphere was slightly hysterical as the clock ticked away and computer equipment behaved in its customary recalcitrant fashion whenever a live demo approaches. Everyone was 'hacking away and hollering and screaming' recalls Vint Cerf, one of the graduate students involved.[9] Kahn himself observed later that if someone had dropped a bomb on the Washington Hilton during the demo it would have wiped out the whole of the US networking community in a single strike.

Anyone who has ever relied on a computer network for a critical presentation knows what a high-risk gamble this was. And yet it paid off: the system worked flawlessly[10] and was seen by thousands of visitors. The Hilton demo was the watershed event that made powerful and influential people in the computing, communica-

tions and defence industries suddenly realise that packet-switching
was not some gimmick dreamed up by off-the-wall hackers with no
telecoms experience, but an operational, immensely powerful,
tangible technology.

So the ARPANET worked. The question for Kahn (and indeed for
the Agency) was: what next? Answer: networking the world. But
how? The ARPANET model was not infinitely extensible for the
simple reason that extending it would have required everyone to
conform to the requirements of the US Department of Defense. And
other people had other ideas. Indeed, even as the ARPANET was
being built, other packet-switched networks had begun to take
shape. The French, for example, had begun work on their Cyclades
network under the direction of a computer scientist called Louis
Pouzin. And Donald Davies's team at the British National Physical
Laboratory had built their own packet-switched network and were
using it every day.

Within the United States too, people had constructed alternative
systems. In 1969, ARPA had funded a project based at the
University of Hawaii, an institution with seven campuses spread
over four islands. Linking them via land-lines was not a feasible
proposition, so Norman Abramson and his colleagues Frank Kuo
and Richard Binder devised a radio-based system called ALOHA.
The basic idea was to use simple radio transmitters (akin to those
used by taxi-cabs) sharing a *common* radio frequency. As with
ARPANET, each station transmitted packets whenever it needed
to. The problem was that because the stations all shared the same
frequency the packets sometimes 'collided' with one another (when
two or more stations happened to transmit at the same time) with
the result that packets became garbled. Abramson and his colleagues got round this by designing a simple protocol: if a transmitting
station failed to receive an acknowledgement of a packet, it
assumed that it had been lost in transmission, waited for a random
period[11] and then retransmitted the packet.

ARPA was interested in this idea of radio links between computers, for obvious reasons. There was a clear military application – an
ALOHA-type system based on radio transmitters fitted in moving

vehicles like tanks could have the kind of resilience that centralised battlefield communications systems lacked. But the limited range of the radios would still pose a problem, necessitating relay stations every few miles – which reintroduced a level of vulnerability into the system. This led to the idea of using satellites as the relay stations, and in the end to the construction of SATNET, a wide-area military network based on satellites. In time, SATNET linked sites in the US with sites in Britain, Norway, Germany and Italy. Another attraction of these systems is that they offered far greater flexibility in the use of shared (not to mention scarce and expensive) transmission capacity.

As these other packet-switching networks developed, Bob Kahn (who had now moved from BBN to ARPA) became increasingly preoccupied with the idea of devising a way in which they and ARPANET could all be interlinked. This was easier said than done. For one thing, the networks differed from one another in important respects. More importantly, they differed greatly in their *reliability*. In the ARPANET, the destination IMP (as distinct from the host computer to which it was interfaced) was responsible for reassembling all the parts of a message when it arrived. IMPs made sure that all the packets got through by means of an elaborate system of hop-by-hop acknowledgements. They also ensured that different messages were kept in order. The basic protocol of the ARPANET – the Network Control Program – was therefore built on the assumption *that the network was reliable*.

This could not hold for the non-ARPA networks, where the governing assumption was, if anything, exactly the opposite. There is always interference in radio transmissions, for example, so the ALOHA system had to be constructed on the assumption that the network was inherently *unreliable*, that one simply couldn't count on a packet getting through. If no acknowledgement was received, a transmitting host would assume it had got lost or garbled and dispatch an identical packet; and it would keep doing this until it received acknowledgement of receipt. Also the non-ARPANET systems differed in other fundamental respects – right down to what they regarded as the size of a standard packet and the fact that they had different rates of transmission. Clearly linking or 'internetting' such a motley bunch was going to be difficult.

To help solve this problem, Kahn turned to a young man who had been with the ARPANET project from the beginning. His name was Vinton ('Vint') Cerf. The partnership between Kahn and Cerf is one of the epic pairings of modern technology, but to their contemporaries they must have seemed an odd couple. Kahn is solid and genial and very middle American – the model for everyone's favourite uncle – who even as a young engineer radiated a kind of relaxed authority. Cerf is his polar opposite – neat, wiry, bearded, cosmopolitan, coiled like a spring. He was born in California, the son of an aerospace executive, and went to school with Steve Crocker, the guy who composed the first RFC. His contemporaries remember him as a slender, intense child whose social skills were honed (not to say necessitated) by his profound deafness. They also remember his dress style – even as a teenager he wore a suit and tie and carried a briefcase. 'I wasn't so interested in differentiating myself from my parents,' he once told a reporter, 'but I wanted to differentiate myself from the rest of my friends just to sort of stick out.' Coming from a highly athletic family, he also stood out as a precocious bookworm who had taught himself computer programming by the end of tenth grade and calculus at the age of thirteen. His obsession with computers was such that Crocker once claimed Cerf masterminded a week-end break-in to the UCLA computer centre simply so that they could use the machines.[12]

Cerf spent the years 1961–5 studying mathematics at Stanford, worked for IBM for a while and then followed Crocker to UCLA, where he wound up as a doctoral student in Leo Kleinrock's lab, the first node on the ARPANET. In this environment Cerf's dapper style set him apart from the prevailing mass of untidy hackers. He was one of the founder members of the Network Working Group, and his name figures prominently in the RFC archive from the begin-ning. He and Kahn had first worked together in early 1970 when Kahn and Dave Walden conducted the tests designed to push the ARPANET to the point of collapse in order to see where its limits lay. 'We struck up a very productive collaboration,' recalled Cerf.

He would ask for software to do something, I would program it overnight, and we would do the tests . . . There were many times when we would crash the network trying to stress it, where it exhibited behavior that Bob Kahn had expected, but that others didn't think could happen. One such behavior was reassembly lock-up. Unless you were careful about how you allocated memory, you could have a bunch of partially assembled messages but no room left to reassemble them, in which case it locked up. People didn't believe it could happen statistically, but it did.[13]

The other reason Kahn wanted Cerf for the internetworking project was because he had been one of the students who had devised the original Network Control Protocol for the ARPANET. Having got him on board, Kahn then set up a meeting of the various researchers involved in the different networks in the US and Europe. Among those present at the first meeting were Donald Davies and Roger Scantlebury from the British National Physical Laboratory, Remi Despres from France, Larry Roberts and Barry Wessler from BBN, Gesualdo LeMoli from Italy, Kjell Samuelson from the Royal Swedish Institute, Peter Kirstein from University College, London, and Louis Pouzin from the French Cyclades project. 'There were a lot of other people,' Cerf recalls, 'at least thirty, all of whom had come to this conference because of a serious academic or business interest in networking. At the conference we formed the International Network Working Group or INWG. Stephen Crocker . . . didn't think he had time to organize the INWG, so he proposed that I do it.'[14]

Having become Chairman of the new group, Cerf took up an Assistant Professorship in Computer Science at Stanford and he and Kahn embarked on a quest for a method of creating seamless connections between different networks. The two men batted the technical issues back and forth between them for some months, and then in the spring of 1973, sitting in the lobby of a San Francisco hotel during a break in a conference he was attending, Cerf had a truly great idea. Instead of trying to reconfigure the networks to conform to some overall specification, why not leave them as they were and simply use computers to act as *gateways* between different systems? To each network, the gateway would look like one of its standard nodes. But in fact what the gateway would be doing was simply taking packets from one network and handing them on to the other.

In digital terms, this was an idea as profound in its implications as the discovery of the structure of the DNA molecule in 1953, and for much the same reasons. James Watson and Francis Crick uncovered a structure which explained how genetic material reproduced itself; Cerf's gateway concept provided a means by which an 'internet' could grow indefinitely because networks of almost any kind could be added willy-nilly to it. All that was required to connect a new network to the 'network of networks' was a computer which could interface between the newcomer and one network which was already connected.

In some ways, the idea of using gateways was analogous to Wesley Clark's 1967 brainwave of using IMPs – the message-switching computers – as intermediaries between host computers and the network. But thereafter the analogy breaks down because ARPANET hosts were absolved of responsibility for ensuring the safe arrival of their messages. All they had to do was to get the packets to the nearest IMP; from then on the sub-network of IMPs assumed responsibility for getting the packets through to their destination. The Network Control Protocol on which the network ran was based on this model. But it was clear to Cerf and Kahn that this would not work for an 'internet'. The gateways could not be expected to take responsibility for end-to-end transmission in a variegated system. That job had to be devolved to the hosts.

And that required a new protocol.

In a remarkable burst of creative collaboration, Cerf and Kahn laid the foundations for the 'network of networks' during the summer and autumn of 1973. In traditional ARPANET fashion, Cerf used his graduate students[15] at Stanford as sounding boards and research assistants, and frequently flew to Washington where he and Kahn burned much midnight oil on the design. In September they took their ideas to a meeting of the INWG at Sussex University in Brighton and refined them in the light of discussions with researchers from Donald Davies's and Louis Pouzin's labs. Then they returned to Washington and hammered out a draft of the scientific paper which was to make them household names in the computing business. It was a joint production, written, Cerf recalled, with 'one of us typing and the other one breathing down his neck, composing as we'd go along, almost like two hands on a pen'. By December, the paper was finished and they tossed a coin to see who would be the lead author. Cerf won – which is why the media now routinely describe him as 'the father of the Internet'.

'A Protocol for Packet Network Interconnection' by Vinton G. Cerf and Robert E. Kahn was published in a prominent engineering journal[16] in May 1974. It put forward two central ideas. One was the notion of a gateway between networks which would understand the protocol used by the hosts that were communicating across the multiple networks. The other was that packets would be encapsulated by the transmitting host in electronic envelopes (christened 'datagrams') and sent to the gateway as end-to-end packets called 'transmission-control-protocol' or TCP messages. In other words, whereas the ARPANET dealt only in packets, an internet would deal with packets enclosed in virtual envelopes.

The gateway, in the Cerf–Kahn scheme, would read only the *envelopes*: the *contents* would be read only by the receiving host. If a sending host did not receive confirmation of receipt of a message, it would retransmit it – and keep doing so until the message got through. The gateways – unlike the IMPs of the ARPANET – would *not* engage in retransmission; they would just pass stuff on. 'We focused on end-to-end reliability,' Cerf said. The motto was 'don't rely on anything inside those nets. The only thing that we ask the net to do is to take this chunk of bits and get it across the network. That's all we ask. Just take this datagram and do your best to deliver it.'[17]

The TCP idea was the electronic equivalent of the containerisation revolution which transformed the transport of international freight. The basic idea in the freight case was agreement on a standard size of container which would fit ships' holds, articulated trucks and rail wagons. Almost anything could be shipped inside a container, and special cranes and handling equipment were created for transferring containers from one transport mode to another. In this analogy, the transport modes (sea, road, rail) correspond to different computer networks; the containers correspond to the TCP envelopes; and the dockside and trackside cranes correspond to the Cert–Kahn gateways. And just as the crane doesn't care what's inside a container, the computer gateway is unconcerned about the contents of the envelope. Its responsibility is to transfer it safely on to the next leg of its journey through Cyberspace.

In July 1975, the ARPANET was transferred by DARPA[18] to the Pentagon's Defense Communications Agency as a going concern. Since the Agency's prime mission was to foster advanced research, not run a network on a day-to-day basis, it had been trying to divest itself of the Net for some time.[19] Having done so, it could concentrate on the next major research task in the area – which it defined as the 'internetting' project.

The first actual specification of the TCP protocol had been published as an Internet Experiment Note[20] in December 1974. Trial implementations of it were begun at three sites – Stanford, BBN and University College, London – so the first efforts at developing the Internet protocols were international from the very beginning.

The earliest demonstrations of TCP in action involved the linking of the ARPANET to packet radio and satellite networks. The first live demo took place in July 1977. A researcher drove a van on the San Francisco Bay-shore Freeway with a packet radio system running on an LSI-11 computer. The packets were routed over the ARPANET to a satellite station, flashed over the Atlantic to Norway and thence via land-line to University College, London. From London they travelled through SATNET across the Atlantic and back into the ARPANET, which then routed them to a computer at the University of Southern California. 'What we were simulating', recalls Cerf,

was someone in a mobile battlefield environment going across a continental network, then across an intercontinental satellite network, and then back into a wireline network to a major computing

resource in national headquarters. Since the Defense Department
was paying for this, we were looking for demonstrations that would
translate to militarily interesting scenarios. So the packets were
traveling 94,000 miles round trip, as opposed to what would have
been an 800-mile round trip directly on the ARPANET. We didn't
lose a bit![21]

The Cerf–Kahn TCP proposal was a great conceptual breakthrough but in itself it was not enough to enable reliable communications between wildly different networks. In fact it took six years
of intensive discussion and experimentation to develop the TCP
concept into the suite of interrelated protocols which now governs
the Internet.

These discussions are chronicled in the archive of Internet
Experiment Notes.[22] The record suggests that the evolution of the
TCP protocol into its present form was driven by two factors – the
intrinsic limitations of the original Cerf–Kahn concept, and the
practical experience of researchers at the Xerox Palo Alto Research
Center (PARC), the lab which Bob Taylor had set up after he left
ARPA and which invented much of the computing technology we
use today – from graphical user interfaces like Microsoft Windows
or that of the Apple Macintosh, to Ethernet local area networking
and laser printing.[23]

The PARC people were deep into networking for the simple
reason that they couldn't avoid it. Having decided years earlier
that computers should have graphic displays (rather than just
displaying characters on a screen) they had faced the problem of
how the contents of a screen could be printed. This they solved by
inventing the laser printer, a wonderful machine which could
translate a pattern of dots on a screen into an equivalent pattern
on paper. But this in turn raised a new problem – how to transmit
the screen pattern to the printer. At a resolution of 600 dots per
inch, for example, it takes something like 33 million bits to
describe a single A4 page! The PARC guys were then placed in the
absurd situation of having a computer which could refresh
(update) a screen display in one second, a printer which could
print the page in two seconds, and a cable between the two which

took nearly fifteen minutes to transfer the screen data for that page from one to the other.

The Ethernet local area networking system was PARC's solution to the transmission problem. It was invented in 1973 by a group led by Bob Metcalfe – the Harvard graduate student who used to test the ARPANET by shipping computer-game graphics across it – and inspired by the ALOHA packet radio system. Like the Hawaiian system, Ethernet used packets to transfer data from one machine to another, and it adapted the same approach to the problem of packet collision: each device on the network listened until the system was quiet, and then dispatched a packet. If the network was busy, the device waited for a random number of milliseconds before trying again. Using these principles, Metcalfe & Co. designed a system that could ship data down a coaxial cable at a rate of 2.67 million bits per second – which meant that the time to transmit an A4 page from computer to printer came down from fifteen minutes to about twelve seconds, and local area networking was born.

With this kind of performance, it was not surprising that PARC rapidly became the most networked lab in the US. Having a fast networking technology meant that one could rethink all kinds of basic computing assumptions. You could, for example, think about *distributed processing* – where your computer subcontracted some of its calculations, say, to a more powerful machine somewhere else on the network. And in their quest to explore the proposition that 'the network *is* the computer', PARC researchers attached all kinds of devices to their Ethernets – from fast printers and 'intelligent' peripherals to slow plotters and dumb printers.

More significantly, as the lab's local area networks proliferated, the researchers had to build gateways between them to ensure that networked resources would be available to everyone. And this in turn meant that they had to address the question of protocols. Their conclusion, reached some time in 1977, was that a single, one-type-fits-all protocol would not work for truly heterogeneous internet-working. Their solution was something called the PARC Universal Packet (forever afterwards known as Pup) which sprang from their need for 'a rich set of layered protocols, allowing different levels of service for different applications. Thus, you could have simple but unreliable datagrams (very useful in some situations), but could also have a higher level of functionality which provided complete error control (but perhaps lower performance).'[24]

As it happens, some people on the INWG were moving towards the same conclusion – that a monolithic TCP protocol which attempted to do everything required for internetworking was an unattainable dream. In July 1977, Cerf invited the PARC team, led by John Shoch, to participate in the Group's discussions. They came with the authority of people who had not only been thinking about internetworking but had actually been doing it for real.

In the end, TCP went through four separate iterations, culminating in a decision to split it into two new protocols: a new *Transmission Control Protocol* and an *Internet Protocol* (IP).

The new TCP handled the breaking up of messages into packets, inserting them in envelopes (to form datagrams), reassembling messages in the correct order at the receiving end, detecting errors and retransmitting anything that got lost.

IP described how to locate a specific computer out of millions of interconnected computers, and defined standards for transmitting messages from one computer to another. IP handled the naming, addressing and routing of packets and shifted the responsibility of error-free transmission from communication links (gateways) to host computers.

The evolving suite (which eventually came to encompass upwards of a hundred detailed protocols) came to be known by the generic term 'TCP/IP'. Other parts of the terminology evolved too: the term 'gateway', for example, eventually came to be reserved for computers which provided bridges between different electronic mail systems, while the machines Cerf–Kahn called gateways came to be called 'routers'. The model for the Internet which they conceived therefore became that of an extensible set of networks linked by routers.

Military interest kept the internetting research going through the late 1970s and into the early 1980s. By that time there were so many military sites on the network, and they were using it so intensively for day-to-day business, that the Pentagon began to worry about the security aspects of a network in which military and scientific traffic travelled under the same protocols. Pressure built up to split the ARPANET into two networks – one (MILNET) exclusively for military use, the other for the original civilian crowd. But because users of both networks would still want to communicate with one another, there would need to be a gateway between the two – which meant that there suddenly was an urgent practical need to implement the new internetworking protocols.

In 1982 it was decided that all nodes connected to the ARPANET would switch from the old NCP to TCP/IP. Since there was some reluctance in some sites to the disruption this would cause, a certain amount of pressure had to be applied. In the middle of 1982, NCP was 'turned off' for a day – which meant that only sites which had converted to the new protocol could communicate. 'This was used', Cerf said, 'to convince people that we were serious.' Some sites remained unconvinced, so in the middle of the autumn NCP was disabled for *two* days. After that, the ARPANET community seems to have been persuaded that the inevitable really was inevitable and on 1 January 1983, NCP was consigned to the dustbin of history. The future belonged to TCP/IP.

Cerf recalls the years 1983–5 as 'a consolidation period'.[25] The great breakthrough came in 1985 when – partly as a result of DARPA pressure – TCP/IP was built into the version of the UNIX operating system (of which more later) developed at the University of California at Berkeley. It was eventually incorporated into the version of UNIX adopted by workstation manufacturers like Sun – which meant that TCP/IP had finally made it to the heart of the operating system which drove most of the computers on which the Internet would eventually run. The Net's digital DNA had finally been slotted into place.

11:
The poor man's ARPANET

Where Plenty smiles – alas! she smiles for few,
And those who taste not, yet behold her store,
Are as the slaves that dig the golden ore,
The wealth around them makes them doubly poor.

George Crabbe, 'The Village', 1783

Despite their informality, the people who created the
ARPANET were an elite group. Ever since Licklider's
time, the prime concern of the Agency had been to identify the
smartest academic researchers in the computer science area and
provide them with the funding needed to pursue their hunches.
ARPA funded the best and the brightest of America's computer
scientists – some of whom built the network, which in turn became
a central facility available to all of Licklider's legatees.

But the ARPA crowd were, by definition, a small minority of the
entire computer science community. There were thousands of
other teachers and researchers all over the United States (and
elsewhere) who yearned to use it – but were excluded from it.
These people understood the significance of what had been created.
They knew as well as anyone that the future of the standalone
computer was about as interesting as that of the standalone
telephone. They passionately wanted to get into networking but
were denied access to its first major manifestation.

And then there were those who had once belonged to the

ARPANET elect but had been obliged to leave it. Many graduate students who had worked on the project, for example, eventually went on to take up teaching and research jobs in labs and universities which were not funded by ARPA. If anything, their sense of frustration and exclusion was even more desperate, for it was worse to have networked and lost than never to have networked at all.

The development of the ARPANET therefore took place against a background of increasing desire for networking in an excluded community which was possessed of great collective expertise. If Uncle Sam was not going to give them access to his precious Net, then they would build their own and to hell with the government. The story of how they did so is a fascinating case study in self-help, determination and serendipity. And it begins with the telephone company which didn't like digital networks.

Bell Labs, the central research and development facility of Ma Bell – the mighty AT&T corporation – is one of those places where lightning strikes not once, not twice but many times. It was here in 1947, for example, that William Shockley, John Bardeen and Walter Brattain invented the transistor – the device which transformed electronics and shaped much of the modern world.

Bell Labs were early and heavy users of computing. In 1957 they found that they needed an *operating system* for their in-house computer centre which – like all computers of its era – was a batch-processing facility. In simple terms, an operating system (or OS in hackerspeak) is the program which runs the computer – it's the software which transforms the machine from a lump of metal and glass into something which can do useful work like word-processing or accessing the Internet or whatever. It's what you see when you first switch on your machine. And whenever the application program you're using (in my case the word-processor on which I'm writing this) needs to do something like save a file or print a page, it does so by communicating with the operating system.

Bell had created an OS called BESYS to run their batch-processing machines and for a time everything was fine and dandy. But by 1964 the Labs were buying much fancier computers and had to decide whether to revise BESYS to cope with this sophisticated kit, or go for something else. Through a decision process which remains unclear to the present day, Bell management decided to team up with General Electric and MIT to create a new and (needless to say) vastly superior operating system. It was to be called MULTICS and would demonstrate that general-purpose, multi-user, time-sharing systems were viable.

MULTICS was a vastly ambitious project, and like all such things ran out of time and over budget. Bell had put two of their best people – Ken Thompson and Dennis Ritchie – on to it, but by 1969 decided that enough was enough and withdrew from the project. 'Even the researchers came to believe', Ritchie wrote later, 'that the promises of MULTICS could be fulfilled only too late and too expensively.'[1] The effort simply went awry, said one of Ritchie's managers, because the partners were naive about how hard it was going to be to create an operating system as ambitious as MULTICS, and because they were too ambitious in what they wanted from the system. It was the problem known to engineers as 'second system syndrome' – the tendency to overload a new system design by putting into it everything you wished had been included in the previous version.[2]

Having pulled out of MULTICS, Bell installed an unexciting but robust OS called GECOS on their computer centre machines. It was not exactly state-of-the-art stuff but it worked and it satisfied the management. To programmers like Thompson and Ritchie, however, withdrawal from MULTICS was a disaster. For one thing, they had been pulled back from an innovative research project. For another, GECOS was not only boring – it was also more or less useless for their primary task, which was software development. So they did what programmers generally do in such circumstances: if the management would not provide them with the tools for the job, they would just have to make their own. They would write a programmer-friendly operating system.

Thompson and Ritchie put up a proposal to purchase a computer which they could use as both a target machine and a test-bed. The management turned down the request, for perfectly good manage-rial reasons: after all there were lots of machines around, running under a perfectly good operating system. So Thompson went rummaging and found an obsolete computer – a DEC PDP-7. It was a tiny machine even in those days: one contemporary rated it somewhere between a Commodore 64 and the IBM PC-AT (the second version of the IBM PC based on the Intel 80286 processor). In today's terms it would be less powerful than so-called 'palm-top' electronic organisers like the Psion Series 5.

Yet on this primitive device Thompson and Ritchie created the most enduring and important operating system ever designed. They called it UNICS as a reaction to MULTICS, but in the end the terminating 'CS' became an 'X' and the system became known as UNIX.[3] Its two creators thought they were simply building a programming environment which would be more appropriate for their work. Little did they know that they were also founding an industry, establishing an academic discipline and launching a secular religion.

UNIX is arguably the most important, and is certainly the most influential, operating system ever conceived. This is partly because of the intrinsic elegance of its design. For one thing, in its early manifestations it was tiny – the UNIX 'kernel' was expressed in only 11,000 lines of code. It's much bigger than that today but still small compared with the 5.6 *million* lines of code which make up Windows NT – the Microsoft operating system designed to provide the same kind of functionality.[4]

Elegance? It may seem odd to non-scientists, but there is an aesthetic in software as there is in every other area of intellectual endeavour. Truly great programmers are like great poets or great mathematicians – they can achieve in a few lines what lesser mortals can only approach in three volumes. Paul Dirac's PhD dissertation ran to only nine pages, but in them was distilled the essence of quantum mechanics. As George Bernard Shaw observed, any fool can write a long book, but it takes real talent to write a good short one. So it is with computer software.

The masterful simplicity of UNIX derives partly from the way it embodies a set of general design principles:

Make each program do one thing well. To do a new job, build afresh rather than complicate old programs by adding new features.

Expect the output of every program to become the input to another, as yet unknown, program.

Design and build software, even operating systems, to be tried early, ideally within weeks. Don't hesitate to throw away the clumsy parts and rebuild them.

Use tools in preference to unskilled help to lighten a programming task, even if you have to make a detour to build the tools, and expect to throw some of them out after you've finished using them.[5]

The second thing that distinguished the Thompson/Ritchie concept was the way it separated out the *kernel* – the core of the operating system – from all the other programs needed to run a computer. Many things which are found in other operating systems are not in the UNIX kernel[6] – they are provided by other programs which use it as a tool. Victor Vyssotsky once wrote that the greatest intellectual achievement implicit in UNIX was Thompson and Ritchie's understanding how much they could leave out of an operating system without impairing its capability. 'To some extent,' he went on, 'that was forced by the fact that they were running on small machines. It may also have been a reaction to the complexity of MULTICS . . . [But] it took some very clear thinking on the part of the creators of UNIX to realize that most of that stuff didn't have anything to do with the operating system and didn't have to be included.'[7]

Separating out the kernel from everything else meant that a standard UNIX release consisted of the kernel plus a set of software tools for doing specific tasks like compiling programs, text editing and so on. This must have seemed awkward at the beginning to anyone accustomed to getting an operating system as a single, sprawling package, but in fact – as we shall see – it was the key to the way in which UNIX eventually became the dominant operating system on the Net.

Thompson and Ritchie never conceived of their system as a product but as an *environment* within which they could do what they did best – write software. Despite that, UNIX diffused quite quickly through Bell Labs and then out into the operational field, where AT&T was making increasing use of minicomputers to manage telephone service, repair and installation tasks. In the early years, none of this came about through any concerted action on the part of management. But, as more and more of AT&T's operational computers came to be UNIX boxes, the company eventually realised that some kind of support system was needed to place the development and maintenance of UNIX on a proper footing. This led to the consolidation of various 'releases' of the system, but it was not until the sixth verion – confusingly christened 'UNIX Timesharing System Sixth Edition V' (usually called 'System Five') – that UNIX reached the state where Bell thought it was ready for wider distribution.

While this was going on, Ritchie had turned his attention to devising a high-level programming language more suited to their needs; by early 1973 he had come up with C^8 – the language which became the lingua franca of applications programming for subsequent decades.

C is a high-level language because it enables the programmer to write instructions in a language which bears a passing approximation to English. Here's an example of a simple program written in C (those who are averse to such things may avert their eyes for a moment):

```
Main()
{
printf("hello, world"\n);
}
```

Now this may not look much like English to you, but compared with the cryptic mnemonics of assembly language or the hexadecimal digit-stream of machine code it is pure poetry to those who understand the lingo. Programmers call it 'source code'. The term 'printf' is a C function (a specially written subprogram) which outputs whatever it finds between the brackets to a printing device.

The characters '\n' instruct the printer to output a new line after printing 'hello, world'.[9]

Computers, of course, cannot understand English, or even pseudo-English of the kind illustrated above. So in order to execute a program written in a high-level language like C, the source code is first of all 'compiled' (that is, automatically translated) into machine code that the computer's processing unit understands, and then submitted to the operating system for execution. This means that a programmer can write in an English-like language and not bother about machine-level code. So long as an appropriate compiler exists for the machine, the appropriate translation will be made.

In its first manifestation, the UNIX kernel was written in *Assembly Language* (or 'Assembler' in hackerspeak), a kind of intermediate language closer to the machine code at which the processor operates and therefore very hardware-specific. Adapting UNIX to run on a new computer meant translating the kernel's 11,000 lines into the assembly language of the new machine – in other words a hell of a business. Some time in 1974, however, Dennis Ritchie solved this problem by rewriting 10,000 of the kernel's 11,000 lines in C. As a result it became relatively simple to 'port' UNIX to (that is, configure the system for) any given type of hardware. Whereas it might take a year to rewrite and debug an assembly-language version for a new processor, reconfiguring the C version took only a matter of weeks. All of a sudden, the most sophisticated operating system in existence had become portable.

Even after the emergence of System V, however, AT&T didn't have a product in UNIX. Or, rather, it did have a product but not one it could sell. In 1949, the US government had filed an anti-trust suit against the company (which, as we saw earlier, had been a lightly regulated monopoly for nearly half a century). After interminable legal arguments, the case was eventually settled in 1956 with the signing of a Consent Decree whereby AT&T agreed to restrict its activities to the regulated business of the national telephone system and government work. One obvious implication of this was that AT&T was barred from entering the computer business. Yet here it was with a sensational operating system on its hands – and one that the computer science community had been itching to get its hands on ever since November 1973, when Ritchie and Thompson had given a presentation at the ACM Symposium on Operating Principles held in Purdue, Indiana.

Faced with the demand for UNIX, and unable to launch it as a commercial product, Bell Labs adopted the policy of effectively giving it away to academic institutions. The cost of a copy of Research UNIX Edition V in December 1974 was $150 – a nominal charge which included a tape of the source code (that is, the C version) and the manuals.[10] By the standards of what commercial operating systems cost at the time, that was dirt cheap.

But its near-zero cost was not why UNIX spread like wildfire through the world's computer science departments. The main reason was that it was the only powerful operating system which could run on the kinds of inexpensive minicomputers university departments could afford. Because the source code was included, *and the AT&T licence included the right to alter the source and share changes with other licensees,* academics could tamper with it at will, tailoring it to the peculiar requirements of their sites. Other attractions were that it was written in a high-level language which was easy to understand, and that it was small enough to be understood in its entirety by students. All of these factors combined to make UNIX the operating system of choice for the majority of computer science departments.

One of the unintended consequences of UNIX was that it gave a powerful boost to the development of computer science as a reputable academic discipline. The system embodied the prevailing mid-1970s philosophy of software design and development. 'Not only was UNIX proof that real software could be built the way many said it could,' wrote John Stoneback,

but it lent credibility to a science that was struggling to establish itself as a science. Faculty could use UNIX and teach about it at the same time. In most respects, the system exemplified good computer science. It provided a clean and powerful user interface and tools that promoted and encouraged the development of software. The fact that it was written in C allowed actual code to be presented and discussed, and made it possible to lift textbook examples into the real world.[11]

And of course the fact that UNIX became ubiquitous in computer science departments meant that it became possible to conduct software 'experiments' which could independently be replicated elsewhere, and of collaboratively building on the earlier work of other researchers – both of which are hallmarks of pukka 'science'.

The traffic in benefits was not all one way, however. UNIX may have been good for computer science, but computer science was also good for it. For one thing, there were all those graduates whose formative programming experiences had been on UNIX machines. For another, many researchers gave something back by writing significant enhancements to the operating system. The *vi* text editor which became part of the UNIX distribution suite, for example, was created by Bill Joy of the University of California at Berkeley. Students at Toronto wrote phototypesetting software for UNIX; Purdue University's Electrical Engineering department made significant enhancements to the UNIX 'shell' (the system's command interpreter); Berkeley students created a new shell; and so on. Many of these improvements and extensions were incorporated in successive upgrades of the system released by Bell. And so the bandwagon rolled on.

From the outset the UNIX phenomenon was intensely collaborative. This was partly a reflection of the fact that the system became so deeply ingrained in academic institutions, and was so infinitely malleable, that there was bound to be a great deal of shared effort and common support among those who relied upon it. Extensive communication between UNIX sites was a necessity, not an option. Operating systems are complex programs and UNIX was no exception. It had bugs which had to be logged, reported, fixed – and the fixes had then to be distributed to users. Thompson and Ritchie imbued licensees with the spirit, 'If something's broken, don't bitch about it – just fix it.' With such a large community of computer-literate users, extensions and improvements were continually being made, and these had to be incorporated in new releases and updated versions – and distributed to the UNIX community.

Meanwhile back at AT&T, as the company came to rely more and more on UNIX, this process of change management had to become formalised. Bugs in the existing code would be discovered and reported to the programmers, or new applications would be created by AT&T departments using the programs for their own tasks. The research labs would need to provide maintenance and updating of software as well as getting the bug reports to the programmer and sending out fixes. In the end, a Bell Labs computer researcher called Mike Lesk proposed an automated maintenance system that would enable the research computer to call up UNIX machines in the various departments, automatically deliver updated software and test that it worked on the remote computers. To implement this idea, he wrote a UNIX program called UUCP (for 'UNIX-to-UNIX copy') which made it possible to have one computer call another computer via a telephone or hard-wired connection and deliver software to it down the line. Lesk could not have known it at the time, but he had invented the charabanc which took the UNIX community on to the information highway.

Being self-evidently useful, a revision of UUCP[12] was duly incorporated in the next release of UNIX and dispatched in late 1979 to the several hundred academic sites then running the system. Among the graduate students who rummaged round in the new release were Tom Truscott and Jim Ellis at Duke University, Durham, North Carolina and Steve Bellovin of the University of North Carolina at Chapel Hill. The three had the idea of using UUCP to get their machines to call one another up using simple, home-made 1,200 bits per second modems, search for changes in specified files held on each machine and then copy the changes over. The process was automated using a program called NetNews, which was essentially a script written in the command language of the UNIX shell.[13]

The system worked reasonably well and soon there were three machines regularly linking in this way – two at Duke and one at UNC. Trouble developed when it became clear that the shell script, which was slow and inefficient in the manner of such things, was tying up valuable machine time, to the annoyance of other users. At this point Truscott and a graduate student called Stephen Daniel rewrote NetNews in C and called the new program 'News version A'.

In January 1980, Jim Ellis described News version A in a presentation at the UNIX Users (USENIX) Association conference. He called it Usenet News. 'The initially most significant service', he predicted,

> will be to provide a rapid access newsletter. Any node can submit an article, which will in due course propagate to all news . . . The first articles will probably concern bug fixes, trouble reports and general cries for help. Certain categories of news, such as 'have/want' articles, may become sufficiently popular as to warrant separate news-groups . . . The mail command provides a convenient means for responding to intriguing articles. In general small groups of users with common interests will use mail to communicate. If the group size grows sufficiently, they will probably start an additional news group . . .[14]

Some months later, the 'A' version of Usenet News software was put on the conference tape for general distribution at the summer USENIX meeting in Delaware. An accompanying handout said that 'A goal of USENET has been to give every UNIX system the opportunity to join and benefit from a computer network (a poor man's ARPANET, if you will) . . .'

Usenet was (and is) what computer folk call a client–server system. You logged on to a machine which in turn was connected to a computer that stored the Usenet postings (articles) for the past few days, weeks or hours. You then examined the headers of articles in the newsgroups to which you 'subscribed' (that is, the ones of interest to you), occasionally requesting the full text of a particular article. The 'client' machine then requested the text from the 'server', which would – if it was still available – deliver it to the user's screen. You then had various options: you could, for example, read the article on screen, print it out or save it to disk; and you could reply to it via electronic mail, send a follow-up article to the same group or start a new subject heading with a new posting.

   Unlike the ARPANET, Usenet News was open to all – or at any rate anyone who had access to a machine running UNIX. Posting and reading news was possible at no cost beyond what universities paid for hardware and the telephone calls to exchange NetNews files.[15] In this way, some of the experiences hitherto available only to the ARPA crowd became available to those without the magic connections or credentials – the 'common people' of the computer science community. Of course Usenet did not enable you to log in as a user on remote machines, or do the other things which were possible on the ARPANET – it merely allowed you to exchange data and information with others in an exceedingly democratic and uncensored way. But, in the end, that was to prove more important than anything else.

   The initial take-off of Usenet News was surprisingly slow. A significant point in its development, however, came when Usenet reached the University of California at Berkeley, which was also an ARPANET node. The Berkeley people created a bridge between the two systems which essentially poured the exchanges in ARPANET discussion groups into Usenet News.[16] This facility highlighted the differences between ARPA and Usenet, for ARPA discussion groups were essentially mailing lists in which the 'owner' of the list decided who was entitlted to receive it, whereas Usenet was constructed on the opposite basis – that individuals decided which News groups they wished to subscribe to. In that sense, Usenet was distinctly more 'democratic' in spirit – which is why it was the model which eventually triumphed on the Internet.

   From a slow start, the 'poor man's ARPANET' grew steadily. Just look at the statistics (see table).[17] In its early years, Usenet articles were mainly transported via UUCP and dial-up telephone lines. As ARPANET opened up, special arrangements were made to allow Usenet to ride on the ARPA network and, later, on the Internet. Its exuberant growth has continued unabated to this day. In 1993, Usenet sites posted about ~26,000 articles per day to 4,902 groups involving the transfer of about sixty-five Megabytes of data around the globe. At the time of writing the number of News groups is somewhere betwen 25,000 and 30,000[18] exchanging God knows how many megabytes a day.

USENET TRAFFIC 1979–1988

Year	Number of sites	Number of articles per day (volume in megabytes)		
1979	3	~2		
1980	15	~10		
1981	150	~20		
1982	400	~50		
1983	600	~120		
1984	900	~225		
1985	1,300	~375	(~1 Megabytes)	
1986	2,500	~500	(2+MB)	
1987	5,000	~1,000	(2.5+ MB)	
1988	11,000	~1,800	(4+MB)	

Usenet has become an integral part of the modern Net. Indeed most Web browsers come complete with an integral 'news reader' which is a direct descendant of the program written by Stephen Daniel and Tom Truscott in 1979.

From the outset, Usenet articles were classified into subject-based categories called news groups, each of which was supposed to limit itself to discussions of a single topic denoted by its name. The groups were in turn organized into *hierarchies* of related topics. Usenet News started with just two hierarchies – 'mod' (for *moderators*) and 'net' (for *network*). The former contained only those news groups which had a gatekeeper who organised material, decided which articles should be 'posted' and generally did the house-keeping. The net hierarchy contained all other news groups.

In 1981 a high-school student called Matt Glickman and a Berkeley graduate student called Mark Horton threw a spanner into these binary works by writing version B of the News software which enabled any Usenet group to be either moderated or open.[19] The release of this new technology and the possibilities it opened up gave rise to an almighty row about authority, categories, hierarchies and free speech generally.

The first outcome was the creation of seven main hierarchies:

Hierarchy	Subject
comp	Computing
misc	Miscellaneous
news	News
rec	Recreational
sci	Science
soc	Society
talk	Anything controversial

Because Usenet worked mainly by dial-up connections, a kind of Usenet 'backbone' had evolved, consisting of those sites which had powerful machines and administrations which were reasonably relaxed about telecommunications charges. These backbone machines subscribed to all newsgroups and acted as local servers for smaller or more impoverished UNIX sites in their vicinity. The managers of these systems gradually came to exercise considerable authority over Usenet and were eventually christened the 'Backbone Cabal'.

Matters came to a head when the user community, continually pushing against the boundaries of the acceptable, wanted Usenet to carry discussions about recreational sex and drugs. A subscriber named Richard Sexton proposed the establishment of groups 'rec.sex'[20] and 'rec.drugs', only to find that the Backbone Cabal refused to carry them. This led to the creation of the 'alt' (for *alternative*) hierarchy of groups which were distributed via communications channels which avoided the backbone and the ARPANET (over which some backbone machines were by this time communicating).

Alt.sex and alt.drugs were the first groups created, on 3 April 1988, followed the next day by alt.rock-n-roll. Brian Reid, who created it, sent the following e-mail to the Backbone Cabal: 'To end the suspense, I have just created alt.sex. That meant that the alt

network now carried alt.sex and alt.drugs. It was therefore artisti-
cally necessary to create alt.rock-n-roll, which I have also done. I
have no idea what sort of traffic it will carry. If the bizzarroids take it
over I will rmgroup it or moderate it; otherwise I will let it be.'[21]

'At the time I sent that message,' Reid later reflected, 'I didn't yet
realise that alt groups were immortal and couldn't be killed by
anyone. In retrospect, this is the joy of the alt network: you create a
group and nobody can kill it. It can only die, when people stop
reading it. No artificial death, only natural death.'[22]

I've just checked the Usenet groups held by my Internet Service
Provider. The alt hierarchy alone boasts 2,521 groups. I scroll down
through them until my thumb cramps on the rollerball I use for
navigating round the Net. And I'm still barely into the alt.c . . .
groups. Here are some of the ones I noted:

Newsgroup	Number of messages currently stored on server
alt.abortion.inequity	665
alt.abuse.recovery	309
alt.adoption	345
alt.algebra.help	58
alt.aliens.imprisoned	7
alt.amazon-women.admirers	287
alt.baldspot	143
alt.banjo.clawhammer	14
alt.bible	458
alt.buddha.short.fat.guy	288
alt.buttered.scones	4
alt.censorship	771
alt.child-support	634
alt.clearing.technology	516

An instructive list, is it not? Three letters of the alphabet and
already we have a compass of the weird diversity of human
interests. The list also gives the lie to one of the nastiest canards
about the Net – that it is a haven for emotionally inadequate nerds

with a predilection for technobabble or New Age psychobabble. True, there are 516 messages from people interested in the technology of clearing banks (though even that is hardly a nerdish preoccupation). And while some of the groups are indeed off the wall (though, interestingly, the subject of imprisoned aliens attracts only seven messages), we also have several hundred messages from souls interested in discussing how people who have been sexually abused can recover from the experience; or over 600 messages about problems of child support.

The evolution of Usenet News illustrates one of the most profound things about the Net, namely that some of its most important benefits have been unanticipated by those who created it. E-mail was a spin-off which came to dominate the ARPANET: yet it took its sponsors by surprise. Usenet News was conceived within the UNIX community as a purely utilitarian device – an effective way of disseminating information about program bugs, fixes and software upgrades through the user community. Yet it turned out to be the medium via which an enormous set of global conversations on an unimaginable range of topics is conducted – and around which a glorious diversity of virtual communities has flourished. 'Think of cyberspace as a social petri dish,' says Howard Rheingold, 'the Net as the agar medium, and virtual communities, in all their diversity, as the colonies of microorganisms that grow in petri dishes. Each of the small colonies of microorganisms – the communities on the Net – is a social experiment that nobody planned but that is happening nevertheless.'[23] Usenet News was the catalyst which made this experiment run.

12:
The Great Unwashed

What man wants is simply independent choice, whatever that independence may cost and wherever it may lead.

Fyodor Dostoevsky, *Notes from Underground*, 1864

In the academic year 1975/6 a colleague of mine went on sabbatical to MIT and returned with tales of people buying strange things called 'personal computers'. We laughed at him, of course, but he countered by producing copies of a magazine called *Popular Electronics* which contained advertisements for, and features about, something called the Altair. Naturally, this made us laugh even more because, by the standards of the DEC minicomputers and mainframes we were using at the time, the Altair was a pretty primitive device. If this was the future then we felt it wouldn't work.

All of which goes some way towards explaining why my net worth today is as close to zero as it can be without actually being bankrupt, while William H. Gates III, Chairman and co-founder of Microsoft Inc., is the richest American in history. For it was that primitive Altair which induced Gates to drop out of Harvard in order not to miss out on what he recognised as the personal computer revolution.

By 1978 there were a lot of Americans around who shared Gates's vision. They were the pioneers of the PC movement – the 'early adopters' so beloved of advertising executives, the guys who simply *have* to have the latest gizmo, the techies who are secretly pleased when the thing doesn't work as advertised because it gives them an excuse to open the hood and poke around. For the most part they were not academics. But many were exceedingly smart cookies who were fascinated by computers, unfazed by the crudity of the early PCs, happy to program in Assembler[1] and generally get their hands dirty. They were also, from the beginning, compulsive communicators – great founders and patrons of 'homebrew' clubs and newsletters and conferences. They constituted the Great Unwashed of the computing community, outsiders who gazed enviously at the ARPANET and Usenet much as poor kids in black neighbourhoods might peer over the wall at rich kids enjoying the facilities of an expensive tennis training complex. Licklider's brats had something these Untouchables wanted – and they were going to get it even if they had to brew it themselves.

The story of the homebrewed Net begins some time in 1977 when two guys named Ward Christensen and Randy Suess decided they wanted to transfer files between their personal computers using the telephone system.[2] Christensen wrote a program called MODEM and released it into the public domain. The basic idea was that you first arranged to have MODEM running on both machines. Then one of you dialled the other and when the phone was answered you both placed the handset in the cradle of a device called an acoustic coupler which enabled the computers to whistle and burble at one another. Once they had established contact in this way, it was possible to transfer files in much the same way as a file could be copied from floppy disk to hard drive on a single machine.

One of the problems with this set-up was the susceptibility of the communications link to interference or *noise* which tended to corrupt the data as they travelled down the line. If you were exchanging text files this might not be too much of a hassle because careful proof-reading was usually sufficient to sort out most garbled passages. But if you were transferring *programs* then even a single missed bit could spell disaster, so in 1979 Christensen and a friend wrote XMODEM, a new version of MODEM which incorporated error correction and which for almost a decade thereafter served as the predominant error-correction protocol for serial-line transmission between computers. Having created it, Christensen again placed the software in the public domain. This had two important effects: first, it meant that anyone seeking to communicate using computers could get the necessary software down the line for free; and secondly, nobody was able to corner the emerging data communications market.

Early in 1978, Christensen upped the ante yet again by writing a program which turned an ordinary PC into a store-and-forward messaging system. He called it 'Computer Bulletin Board System' (CBBS). While he was doing this, Suess designed the hardware for a simple microcomputer-to-microcomputer communications system. The two described the system in the November issue of *Byte*, which by then was already the magazine of record for the digital cognoscenti. Thus the pioneers not only created a new technology – they told everybody else how to do it!

It's important to realise how genuinely low-tech BBS technology was (and is). 'For less than the cost of a shotgun,' Howard Rheingold observed, 'a BBS turns an ordinary person anywhere in the world into a publisher, an eyewitness reporter, an advocate, an organizer, a student or teacher, and potential participant in a worldwide citizen-to-citizen conversation.'[3] To run your own Bulletin Board you did not have to possess a fancy UNIX machine or have access to the ARPANET. All you needed was an ordinary PC, some free software, a slow modem and a telephone line. Once you had the software running the only things that remained to be done were: decide on a name for your BBS; plug the modem into the phone jack; post the telephone number on a few existing BBSs; wait for people to call and 'post' their private messages or public information on your virtual noticeboard and – hey presto! – you were suddenly in the networking and virtual community businesses. And all without anyone at the Pentagon or the phone company having the slightest inkling of what you were up to.

One of the great things about running a BBS was that you were lord of all you surveyed. You could use one, wrote Rheingold, 'to organise a movement, run a business, coordinate a political campaign, find an audience for your art or political rants or religious sermons, and assemble with like-minded souls to discuss matters of mutual interest. You can let the callers create the place themselves, or you can run it like your own private fiefdom.'[4]

The downside of this freedom was, of course, that many BBSs were puerile, racist, sexist, obscurantist or just plain weird. But whatever they were they all shared one important characteristic: the only way to stamp them out would be simultaneously to outlaw the telephone system and the digital computer, and not even the federal government was likely to contemplate that.

The first BBS went online in Chicago some time in 1979. It enabled people to dial in and leave messages in a virtual public 'space', much as people do on notice boards in the foyers and hallways of university departments. The technology spread rapidly and BBS systems quickly appeared in other US cities, each with their coterie of dial-up users, members, subscribers or whatever. But, although the number of systems increased rapidly, each one was, in a sense, a little island in Cyberspace, complete unto itself.

The man who changed all that was a striking individualist called Tom Jennings. He was a programmer in a small software company in Boston, and in 1980 he started using an acoustic coupler to call up the Chicago CBBS. In 1983, he moved to California and found himself with time on his hands while waiting to start work. So he wrote his own BBS program and called it Fido. Why Fido? Well, because he had once homebrewed a computer system out of disparate parts liberated from other machines. The result was such an electronic mongrel that someone had christened it Fido and the name stuck.[5]

What interested Jennings was not so much the challenge of creating a standalone BBS, as it were, but of creating a *network* based on BBS technology. By December 1983 he had Fidonet node #1 online in the Bay Area. The next year Jennings helped John Madill in Baltimore set up Fidonet node #2, after which the network began to spread with great rapidity because the software to run a node was freely available on the network itself. By the end of 1984, between thirty and forty nodes were operational.[6] By the end of 1986 there were over a thousand. Rheingold estimated that on the (conservative) assumption of ten users per node, about 10,000 people were using this homebrewed Net. By 1991 there were over 10,000 nodes, all over the world – and therefore about 100,000 users. In other words, this unofficial, anarchic, amateurish system had a user population not that far removed from that of the Internet at the time.[7]

Fidonet was an authentic store-and-forward network. Each node would accumulate a day's messages, sort out which had to be passed on to other nodes and then dial the target node to deliver its messages. Jennings's original dream was that everything would travel free on his net, exploiting the fact that most local telephone calls in the US were free. (This was largely a consequence of AT&T's monopoly; charges on long-distance calls subsidised local ones.) Individual Fidonet nodes would call nearby ones toll-free, pass on messages and hang up. The recipients would call some more nodes, again toll-free, and pass on the mail. In this way, free transcontinental electronic messaging would be possible.

It was a crazy idea, of course, and Jennings dropped it when he realised that if the nodes dialled in the middle of the night, when long-distance charges were lowest, it would be more efficient to send stuff direct. This was the beginning of 'National Fido Hour': between the hours of 1 and 2 a.m. every night, Fidonet nodes spurned their dial-up users, called one another and exchanged messages. Jennings gave each node a unique network address; people would put the destination node address on each message and the network would forward it until it arrived at the node for which it was intended. The whole thing was organised round a node-list maintained by Jennings and distributed via e-mail. As nodes began to proliferate in major cities, he also designated some as local 'gateways'. So instead of making thirteen separate calls to computers in the Chicago area, say, a node in San Francisco would send the messages in a stream to the Chicago gateway, which would then route them to local machines.

Later on, Fidonet was able to use the Internet for long-haul connections. A message originating in Carmel, California, for example, and destined for someone in Aachen, Germany, might travel via a dial-up link to a node in San Franciso which happens also to be an Internet gateway. It then rides at lightspeed across the Atlantic to Amsterdam, where it resumes its dial-up hopping until it winds up on the destination node in Aachen. It may sound crazy and look inefficient, but it is as much in the spirit of Baran's concept of a resilient network as the ARPANET itself.

As I write, Fidonet is still going strong, with nearly 40,000 nodes and an estimated 3 million users worldwide.[8] It has six continental zones and provides electronic mail, online conferencing and file distribution services. It reaches into areas of the world – notably Africa – which the Internet passes by. But the most astonishing thing about Fidonet is the fact that it is powered by nothing more tangible than enthusiasm – whatever it is that motivates someone to make his or her computer available to others and join in a global, co-operative enterprise whose mission is simply to enable people to exchange messages and other forms of digital data. Anyone can set up a Fidonet node.[9] Network people sometimes sneer that Fidonet is an 'amateur operation', but in doing so they pay it a back-handed, unintended compliment. For while it is undoubtedly true that most of the 'Sysops' ('System Operators'), who run Fidonet nodes are not network professionals, they know their technical stuff. In fact they *are* amateurs in the original and best sense of the term: 'amateur' comes from the Latin verb *amare*, to love. Fido Sysops do it for love, or at least for the hell of it. And they have the satisfaction of knowing that if the entire Internet turned to jelly tomorrow, their homebrewed Net would still get messages through to their destinations. And that there is no power on earth that could kill it. At a time when governments and multinational corporations are itching to get the Internet under (their) control, it's deeply reassuring to know that the framework for an alternative, free communications system not only exists, but thrives.

What did Usenet and the Great Unwashed bring to the networking party? Ask James Exon.

In 1995 Mr Exon, then an outgoing Senator from Nebraska, introduced a Bill into the US Congress designed to censor the Internet. At a family gathering the previous Christmas, he had been horrified to discover that some of his grandchildren and their friends had access to the Net – and he had been even more appalled to discover that there was *pornography* on the Web! Gadzooks! The

Bill was his retirement gift to a dozy Congress, and on 8 February 1996 President Clinton signed it into US law as the Communications Decency Act.

The nub of the CDA appears in two parts of Section 223. One provides that any person in interstate or foreign communications who, 'by means of a telecommunications device knowingly makes, creates, or solicits' or 'initiates the transmission of any comment, request, suggestion, proposal, image or other communication which is obscene or indecent, knowing that the recipient of the communication is under 18 years of age' shall be guilty of a criminal offence and, on conviction, fined or imprisoned. The second makes it a crime to use an 'interactive computer service' to 'send' or 'display in a manner available' to a person under age eighteen, the following material: 'any comment, request, suggestion, proposal, image, or other communication that, in context, depicts or describes, in terms patently offensive as measured by contemporary community standards, sexual or excretory activities or organs, regardless of whether the user of such service placed the call or initiated the communication'.

These unworkable and sweeping infringements of freedom of speech were challenged in June 1996 by the American Library Association, the American Civil Liberties Union and others in a US district court in the Eastern District of Pennsylvania. The case was heard by three judges – Dolores Sloviter, Ronald Buckwalter and Stewart Dalzell – who found that the two subsections complained of did indeed violate the First and Fifth Amendments to the US Constitution. It was a landmark judgment[10] not just because of the result, but also because of the astonishing way in which the three judges had penetrated to the heart of the Net and understood its essence.

'It is no exaggeration', wrote Judge Dalzell in his summing-up,

> to conclude that the Internet has achieved, and continues to achieve, the most participatory marketplace of mass speech that this country – and indeed the world – has yet seen. The plaintiffs in these actions correctly describe the 'democratizing' effects of Internet communication: individual citizens of limited means can speak to a worldwide

audience on issues of concern to them. Federalists and Anti-Federalists may debate the structure of their government nightly, but these debates occur in newsgroups or chat rooms rather than in pamphlets. Modern-day Luthers still post their theses, but to electronic bulletin boards rather than the door of the Wittenberg Schlosskirche. More mundane (but, from a constitutional perspective, equally important) dialogue occurs between aspiring artists, or French cooks, or dog lovers, or fly fishermen.

Dalzell went on to ridicule the pretensions of the CDA by seeking analogies to it in the print world. 'I have no doubt', he said,

that a ban on a front page article in the New York Times on female genital mutilation in Africa, would be unconstitutional . . . Nor would a Novel Decency Act, adopted after legislators had seen too many pot-boilers in convenience store book racks, pass constitutional muster . . . There is no question that a Village Green Decency Act, the fruit of a Senator's overhearing of a ribald conversation between two adolescent boys on a park bench, would be unconstitutional . . . A Postal Decency Act, passed because of constituent complaints about unsolicited lingerie catalogues, would also be unconstitutional . . . In these forms of communication, regulations on the basis of decency simply would not survive First Amendment scrutiny.

The Internet is a far more speech-enhancing medium than print, the village green, or the mails. Because it would necessarily affect the Internet itself, the CDA would necessarily reduce the speech available for adults on the medium. This is a constitutionally intolerable result.

In coming to this conclusion, Judge Dalzell was under no illusions about what freedom of speech means on the Net. 'Some of the dialogue on the Internet', he wrote, 'surely tests the limits of conventional discourse.'

Speech on the Internet can be unfiltered, unpolished, and unconventional, even emotionally charged, sexually explicit, and vulgar – in a word, 'indecent' in many communities. But we should expect such speech to occur in a medium in which citizens from all walks of life have a voice. We should also protect the autonomy that such a medium confers to ordinary people as well as media magnates. Cutting through the acronyms and argot that littered the hearing testimony, the Internet may fairly be regarded as a never-ending worldwide conversation. The Government may not, through the CDA, interrupt that conversation. As the most participatory form of mass speech yet developed, the Internet deserves the highest protection from governmental intrusion.

The government, of course, appealed to the Supreme Court. But on 26 June 1997 that court agreed with Judge Dalzell and his colleagues that the Net – however raw, unfiltered, explicit, vulgar, decent or indecent it may be – is entitled to the protection of the First and Fifth Amendments.[11]

The point of this little story is not to highlight the advantages of an independent judiciary or the stupidity of Senators but the way the Net has become a Jeffersonian market-place in ideas. We have come a long way – and not just in chronological time – from the ARPANET. The part of Cyberspace defined by the Pentagon's experimental system was a pretty refined place by the standards of today's Net. Sure, it had some discussion groups organised round mailing lists, and sure they discussed some ostensibly way-out topics like science fiction and human behaviour, but in the main the ARPA crowd were pretty focused on engineering and computer science.

It was the outsiders – the Usenet crowd and Tom Jennings's Great Unwashed – who made the Net the unruly place it is today. It was they, for example, who introduced the idea that networks were for conversation more than for file transfer. It was they who established the principle that the proper subject of computer-mediated conversations were anything that human beings wanted to discuss. It was they who introduced the technology of asynchronous conferencing via news groups.[12] The original ARPANET community was a disciplined, orderly, relatively homogeneous, anal-retentive group which would not have offended Senator Exon. Their descendants are brash and undisciplined and – praise the Lord – beyond anyone's control.

13:
The gift economy

God answers sharp and sudden on some prayers,
And thrusts the thing we have prayed for in our face,
A gauntlet with a gift in't.

Elizabeth Barrett Browning, 'Aurora Leigh', 1857

In the summer of 1956, a group of young researchers gathered at Dartmouth College, New Hampshire to discuss a revolutionary idea – the notion that something called 'artificial intelligence' might be possible. The conference was funded by the Rockefeller Foundation and was organised by John McCarthy, Marvin Minsky, Nathaniel Rochester and Claude Shannon. Among the other participants were Allen Newell and Herbert Simon (who later went on to win a Nobel Prize for economics).

It's not often one can pinpoint the precise point at which an academic discipline is born, but the Dartmouth conference was such a moment. Minsky defined the new field as 'the science of making machines do things which would require intelligence if done by a human' and from this a distinctive way of doing AI research rapidly emerged. It went something like this: select a human behaviour or facility which, by common consent, requires 'intelligence' (playing chess, say, or solving differential equations); write a computer program which can display, to some extent, the same behaviour or facility; and evaluate the extent to which the program achieves that objective. In a way, AI was conceived as a counterpoint to conventional cognitive science. But unlike psychology it regarded the processes of human cognition as effectively inaccessible. By creating computer programs which could behave intelligently, AI researchers believed they could shed light on human intelligence as well as its artificial counterpart.

The underlying assumption of the pioneers of AI was that the things humans find difficult machines would find difficult and vice versa. In the event, they discovered that the opposite seems to be true. It proved unexpectedly easy to make computers do what many people find difficult (algebra, calculus, chess, for example). But human abilities possessed by any three-year-old child (three-dimensional vision, natural language, remarkable sensory-motor co-ordination) seemed to lie way beyond the capabilities of machines. In the end, this ironic paradox caused a crisis in the discipline from which it's still struggling to recover.[1]

But this is all with the twenty–twenty vision of hindsight. Way back in the late 1950s, one thing that was clear to McCarthy, Minsky & Co. was that AI research would require computing – a lot of computing, and sometimes very heavy-duty computing. From the outset, AI researchers were probably the most demanding users of computers after the US military. They spent their time writing very complex programs, and they needed languages and programming environments, tools and specialist hardware to make them more productive. For example, John McCarthy, who invented the high-level LISP language for AI work in the late 1950s, was – as we saw in an earlier chapter – one of the pioneering advocates of time-sharing.

In 1960, an Artificial Intelligence Project was set up at MIT under the direction of McCarthy and Minsky. Much of its funding came, needless to say, from ARPA. Ten years later the 'project' had matured into the MIT Artificial Intelligence Laboratory under the direction of Minsky and Seymour Papert, a South African visionary who had studied with Jean Piaget, invented the Logo programming language and campaigned passionately for the view that computers could be used to liberate and harness the learning abilities of children.[2]

The AI Lab at MIT was a geek's paradise. It was lavishly funded, attracted terrific people and fascinated the computer industry and the media. There was an intellectual buzz about the place. Colleagues of mine who visited the Lab would come away shaking their heads in awe at what they had seen – smart people working on way-out projects; fantastic ideas for robotics research; and the kind of kit the rest of us would kill for – specialised, individual workstations fitted with umpteen megabytes of RAM, huge hard disks and twenty-one-inch screens (at a time when the rest of us were making do with 640k of RAM, ten-megabyte hard drives and character-based displays).

One of the AI Lab's best-known members is a guy named Richard Stallman, who joined in 1971 as an eighteen-year-old Harvard student. He is largely unknown in the outside world, but within the programming community Stallman enjoys a status somewhere between Superman and Moses. When you see him in the flesh, it's easy to understand the Biblical association, for with his long, stringy hair, piercing green eyes and utter indifference to the opinions of others, he looks the part of an Old Testament prophet. 'I feared his wrath,' wrote one interviewer. 'As he pointed out to me repeatedly through the course of our afternoon together, he doesn't do things because they are socially acceptable or strategically appropriate. Success is not his measure for accomplishment. He does what he does because he thinks it is the morally correct, or simply fun, thing to do. And he brooks no compromise.'[3] Stallman is revered by many on technical grounds and admired by some on moral ones. Hackers know him as the creator of the EMACS editor, a remarkably sophisticated and versatile editing program which runs under UNIX and has been the basic working tool of hundreds of thousands of sophisticated programmers over the years. Those who regard him as a saint venerate him for his passionate, all-consuming conviction that software should be free – and for his magnificent disdain for the values which have made Bill Gates the richest man on earth.

Stallman did not always think like this. In fact, when he started in computing, it probably never occurred to him – or indeed to many of his colleagues – that software would be anything other than free. His professional socialisation took place in a research lab where code was co-operatively written, freely shared and always regarded as being in the public domain – in exactly the same way as scientific publications are, in fact. The culture in which Stallman grew up was that of the graduate students who developed the ARPANET protocols, a culture which had no professional secrets, in which co-operative effort was the order of the day and in which the only judgement worth bothering about was that of one's peers.

Two events shocked Stallman into a realisation that this culture was threatened. The first came one day in 1979 after Xerox had donated one of the first laser printers to the AI Lab. The printer crashed a lot so Stallman decided to fix it. To do that he needed to have a copy of the source code – the high-level version of the program which ran the printer. His idea was to modify the program so that the printer would respond to breakdowns by flashing a warning on to the screens of everyone who had sent it a print job and was waiting patiently for it to be done. Given that all these people had an incentive to reset the printer after a crash, Stallman figured this would be a neat way of reducing printer downtime.[4]

So he did what programmers then did all the time – requested a copy of the source code. He had done this once before for an earlier, equally problematic printer, and Xerox had obliged. But this time the company refused to supply the source. It was, Xerox told him, now copyrighted material. He was not allowed access to it, even if his intention was to improve it.

Stallman was enraged, but this in itself was not enough to trigger him to do something about what he saw as the underlying problem. The final straw came a couple of years later when many of his colleagues in the AI Lab left to found Symbolics, a company devoted to building and selling high-powered AI workstations. Stallman was appalled at the way his professional peers, the researchers with whom he had worked openly and co-operatively for so long, were suddenly transformed into secretive guardians of intellectual property. 'When I saw the prospect of living the way the rest of the world was living,' he told Andrew Leonard, 'I decided no way, that's disgusting, I'd be ashamed of myself. If I contributed to the upkeep of that other proprietary software way of life, I'd feel I was making the world ugly for pay.'[5]

In 1984, Stallman launched the Free Software Foundation[6] to campaign for his vision of how the world ought to be. The title of the foundation has, over the years, puzzled people, especially journalists. Does Stallman really believe that even the most complex programs – the ones that have taken tens of thousands of programmer-hours to create – should be just *given away*? In such a world, how would anyone make a buck, let alone a living?

But this is not at all what Stallman had in mind when he set up his foundation. 'Free software', he wrote, 'is a matter of liberty, not price. To understand the concept, you should think of "free speech", not "free beer".' What he means by free software is the *user's* freedom to run, copy, distribute, study, change and improve the software. This freedom should exist, he believes, at three levels. First, you should be able to study how the program works, and to adapt it to your needs. Secondly, you should be able to redistribute copies so you can share with your neighbours/peers/collaborators. And thirdly, you should be free to improve the program and release your improvements to the public so that the whole community benefits.[7]

The problem with implementing such an idealistic scheme, of course, is how to stop unscrupulous people who do not share your altruistic values from ripping you – and the community – off. The simplest way to make a program free is to put it in the public domain, uncopyrighted. This allows people to share and improve the software, as Stallman wanted. But it also allows shysters to convert the program into proprietary software. What's to stop them making a few small changes and distributing the altered software as their own intellectual property?

Stallman's great idea was to devise a licensing system which would prevent this happening while enabling all the good things he sought. Instead of releasing free software into the public domain, he released it with a licence which required that anyone who redistributed it, with or without changes, must pass along the freedom to further copy and change it. And he called this new licensing system *copyleft* to distinguish it from traditional copyright.

Having sorted out the licensing issue. Stallman than turned his attention to finding software which would be worth liberating. And as it happened, the US Department of Justice dropped it into his lap. It was called UNIX.

UNIX, as we have seen, was the property of AT&T, in whose research labs it had been created, but the company was debarred from selling it because of the 1956 Consent Decree which forbade it from entering the computer business. As a result, AT&T had for years effectively been giving UNIX away to academic institutions and research labs under licensing terms which bore some resemblance to those desired by Stallman. But in 1974 the US government, concerned that changes in technology were making the continuation of AT&T's monopoly indefensible, filed another anti-trust suit against the company. As in the previous (1949) case, the legal arguments dragged on for years, but a deal was finally reached in 1982 when AT&T agreed to divest itself of its wholly owned Bell operating companies which provided Americans with local telephone services. In return, the US Department of Justice agreed to lift the constraints of the 1956 decree.

The old, monopolistic Ma Bell was broken up on 1 January 1984 and thereafter AT&T was free to capitalise on its ownership of UNIX.[8] Suddenly the operating system which generations of computer science students and hackers had taken for granted became a product just like any other. It was also priced accordingly – and operating systems never came cheap. And the freedom to tamper with the system's source code – and redistribute the results – which graduate students had taken as their birthright was suddenly taken away. It was a disaster, an outrage, a scandal.

To Richard Stallman, however, it was the challenge he had been born for. As the creator of EMACS, the editing tool once described as 'a kind of nuclear-powered Swiss Army knife for programmers',[9] he knew as much about the innards of UNIX as any man living. The idea that this wonderful operating system should become a commercial product just like anything sold by Microsoft struck him as too sordid for words. So he decided to do something about it. In fact he decided to create, from scratch, a clone of AT&T UNIX which would not infringe the company's copyright, but which would be indistinguishable to users from the real thing. In one of those recursive jokes beloved of hackers,[10] he called his new system GNU (pronounced 'Guh-new'). It stands for 'Gnu's not UNIX'. From that moment on, Stallman's quixotic enterprise has been universally known as the GNU Project.

The task he had set himself was stupendous. An operating system is a very complicated program. It has to be because it provides all the services that application programs need. Whenever a program like Microsoft Word wants to move a file on to a disk, for example, or move a symbol from the keyboard into RAM, or send a file to the printer, or do a thousand and one other things, it has to ask the operating system to intervene on its behalf. To provide such a level of service, the operating system has to have hundreds, maybe thousands, of subsidiary programs to perform these myriad specialised tasks. To create a functioning clone of 1980s UNIX, Stallman would have to write not only a new kernel, but all these subprograms as well. It would be, observed one of his former colleagues, 'like building a jet plane from scratch in your garage. People thought it was impossible. And it probably would have been, if anyone less extraordinarily talented than Richard was in charge.'[11]

Stallman and a group of dedicated hackers worked like crazy for years on the GNU project. Stallman himself left the Lab only rarely, preferring to sleep on a folding bed in his office until the MIT authorities prevailed upon him to doss down elsewhere. Eventually his hands, wearied by interminable typing, wore out; acute pain prevented him from using a keyboard, and his work on the kernel stopped. For a time he tried to continue by employing MIT undergraduates as transcribers. 'He would treat them literally as typewriters,' a colleague recalled, 'saying "carriage return" and "space" and "tab", while he dictated what he saw in his head.'[12] But invariably his assistants, demoralised by days of robotically transcribing computer code, would quit, leaving Stallman stranded.

In the course of their interminable death-march through the 1980s the GNU hackers produced scores of the subsidiary programs needed by the operating system – and distributed them all over the world. Naturally, every line they wrote was 'copylefted'. But they never managed to produce the heart of the system – a new kernel. And without that they couldn't have a working clone of the

wonderful system that Thompson and Ritchie had created and
AT&T was now exploiting. It looked as though the forces of
copyright had won.

They might have done if an undergraduate named Linus Torvalds
had not bought himself a PC in 1991. At the time this soft-spoken,
sandy-haired, understated Finn was a twenty-one-year-old student
at the University of Helsinki. He had spent some of his adolescent
years writing games programs for early home computers and had
been particularly struck by the Sinclair QL, an eccentric British
machine which had many peculiarities but one amazing feature – a
genuinely multi-tasking operating system that allowed one to do
serious programming.

In the autumn of 1990, Torvalds enrolled on a UNIX course and
was given access to a DEC MicroVAX running DEC's particular
flavour of the operating system. The problem was that the machine
could only cope with sixteen users at a time, and he got fed up
waiting in line for access to a terminal. Among his textbooks for the
course, however, was Andrew Tanenbaum's *Operating Systems:
Design and Implementation*,[13] which described a small, but real
UNIX-like operating system: MINIX. The book demonstrated how
it worked while illustrating the principles behind it. It also provided
the MINIX source code.

When it first came out in 1987, MINIX caused something of a
sensation. Within weeks, it had its own news group on Usenet, with
40,000 people – many of whom badgered Tanenbaum to make it
more elaborate by adding features.[14] Since what he wanted to do
was to create something small enough to be understood by the
average undergraduate, Tanenbaum wisely declined these requests.
But, in doing so, he unwittingly left the door open for Torvalds.

MINIX was what persuaded the young Finn that the time had
come to get his own PC. Until then he had resisted, he said,
because 'if I had gotten a PC, I'd have gotten this crummy
architecture with this crummy MS-DOS operating system and I
wouldn't have learned a thing'.[15] With his new PC, Torvalds began
to experiment, using MINIX as scaffolding to develop a new
program. He says that he never intended to create a kernel. Instead,
a purely practical need to read Usenet news groups drove him to

modify some basic MINIX processes to enable him to do that. Then he found he needed something else: software drivers – the programs which drive peripheral devices like screen, printers, keyboards and modems. A driver acts as an intermediary between the processor and something peripheral (like a modem). So Torvalds wrote some drivers.

And that's how he went on. In the summer of 1991 – just six months after he got his first PC – he found he needed to download some files from the Net. But before he could read and write to a disk, he recalled, 'I had to write a disk driver. Then I had to write a file system so I could read the Minix file system in order to be able to write files and read files to upload them . . . When you have task-switching, a file system, and device drivers, that's Unix.'[16] Or at least its core. In this way he more or less backed into one of the seminal computing achievements of the century – a UNIX-like operating system which did not belong to AT&T, or indeed to anyone else. This kind of thing probably happens all the time in computing, and Torvalds's operating system might have remained an obscure effort by an unknown Scandinavian student had not its creator posted a message about it in the MINIX news group, comp.os.minix. It was dated 25 August 1991 and read, in part:

Hello everybody out there using minix –

I'm doing a (free) operating system (just a hobby, won't be big and professional like gnu) for 386(486) AT clones. This has been brewing since april, and is starting to get ready. I'd like any feedback on things people like/dislike in minix, as my OS resembles it somewhat (same physical layout of the file-system (due to practical reasons) among other things).

I've currently ported bash (1.08) and gcc(1.40),[17] and things seem to work.

This implies that I'll get something practical within a few months, and I'd like to know what features most people would want. Any suggestions are welcome, but I won't promise I'll implement them:-)

Linus (torvalds@kruuna.helsinki.fi)

PS. Yes – it's free of any minix code, and it has a multi-threaded fs. It is NOT protable [sic] (uses 386 task switching etc), and it probably never will support anything other than AT-harddisks, as that's all I have:-(.[18]

The message prompted an offer of space on a server at the Helsinki University of Technology, letting people download the first public version of the system. One of the biggest problems Torvalds had was deciding on a name for his baby. His working name for it was Linux (pronounced 'linn-ux'), but he was afraid that if he called it that people would think he was an ego-maniac, so he called it 'Freax' – which came from some weird conjunction of 'free', 'freak' and 'x'. It was a dumb idea, but fortunately the guy who ran the server didn't like the Freax label, so he used the working name instead.

From the outset, Linux was copylefted software.[19] Ten people downloaded the first posted version of the program, and five sent back bug fixes, code improvements and new features. By December 1991 about a hundred people were participating in Usenet discussions about Linux and hacking the code. A year later, there was a fully functional Linux operating system running on the IBM PC family of machines. By 1993 there were about 20,000 Linux users worldwide and about 100 programmers actively contributing changes to the code. By 1994 there were 100,000 users and networking facilities had been added to the system, which by now ran to 170,000 lines of code. And so it went on, until today there are somewhere between 7.5 million and 10 million users and upwards of 10,000 programmers actively involved in Linux news groups, testing and code improvements.[20]

And now here's the astonishing bit: in this maelstrom of co-operative development, there is order and progress. The reason Linux has become so popular is not because it makes people feel good to be using something not made by Microsoft (though of course that helps in some cases), but because it is a remarkably stable and robust product. If I had to bet my life on an operating system tomorrow, I would choose Linux. My personal Web server runs under it and when I last checked the machine had been running continuously for nearly a year without once needing to be rebooted. Compare that with my Windows 95 desktop machine, which starts to degrade after a single day's intensive use. And the reason for Linux's astonishing performance is obvious: it is because it has been debugged and tested to destruction by a larger army of skilled programmers than Microsoft could ever assemble.[21] It's much the same process as the one on which we rely to produce scientific knowledge. And because of Stallman's copyleft concept, each participant in the process feels they have a stake in improving the system. The story of Linux is one of those rare cases where the right thing to do turned out to be the right thing to do.

What's made this extraordinary phenomenon possible? Answer: a combination of three things – the copyleft licensing system and the willingness to share source code that it engenders; the Net; and the distinctive mindset of those who work on the Linux kernel.

Source code: commercial (copyrighted) programs are never distributed as source code but as what are called *binaries*. As a customer (or user) you never get to see what the programmer wrote, only the translation of it into incomprehensible binary code. A prerequisite for the Linux and GNU projects is that programmers make the source code of what they produce freely available to others. That is why the Linux and GNU projects are nowadays often described as part of the *Open Source* movement.[22]

The Net: enables programmers working in the Open Source tradition to work collaboratively at a distance. Many of those who contribute to these projects are programmers and system managers on machines scattered all over the world. Without a communications medium which enables them to converse via e-mail and online conferences, and to ship program files back and forth around the world, the pace of development on GNU–Linux would have been much slower. As it is, the Open Source movement makes the pace of traditional software development look positively arthritic. Whereas new versions of major software packages are issued once a year (or once every three years in the case of Microsoft Windows), for example, upgraded versions of Open Source programs often appear monthly.

The mindset: This is the most astonishing aspect of the GNU–

Linux phenomenon. Computer programmers tend, by and large, to be quirky and highly individualistic. Trying to organise or manage such awkward characters is normally as thankless as herding cats. Indeed often the only way companies like Microsoft can entice and manage gifted programmers is by providing serious financial inducements in the form of stock options and other occupational perks.

But these strategies are not available to the Open Source movement. For one thing, it has no money. Linux is available, free of charge, to anyone who wants it and can find a download site. Or it's available in packaged distributions like Red Hat's[23] which cost less than $50 and include manuals and installers. When IBM (which even today is a $100 billion company) tried to strike a deal with the group of Open Source hackers who created the Apache Web-server program, the company's lawyers were baffled to discover that there seemed to be nobody to whom the company could pay a licence fee – the group had no legal existence. Even more puzzling was the fact that the twenty or so hackers involved in the Apache project seemed completely uninterested in money. Their only concern was that any deal with IBM would have to respect the copyleft principle.

Why did IBM want Apache? Simple: it's the world's best program for hosting Web sites, period. About half of all the Web servers in the world are powered by Apache. And most of the mission-critical ones – including Hotmail, the Web-based e-mail system now owned by Microsoft – are Apache sites. Yet this remarkable program, which has 50 per cent of the global market, was created by a group of programmers scattered round the world who rarely meet and derive no serious financial benefit from it.

What motivates such people? A clue emerges from the IBM–Apache story because in the end a deal *was* struck: Apache allowed IBM to incorporate their software as the cornerstone of an online commerce system it was building; in return the company provided the Apache group not with cash but with *an improvement on their software* which had been worked out by IBM programmers – an adjustment which made Apache work better when running on Microsoft NT servers![24] In other words, what persuaded the Apache crowd to agree was the gift of what they call a 'neat hack' – that is, a smart piece of programming.

The IBM lawyers were no doubt as baffled by this as they would have been by a potlatch ceremony in some exotic tribe. But to those who understand the Open Source culture it is blindingly obvious what was going on. For this is pre-eminently a high-tech *gift economy*, with completely different tokens of value from those of the monetary economy in which IBM and Microsoft and Oracle and General Motors exist.

'Gift cultures', writes Eric S. Raymond, the man who understands the Open Source phenomenon better than most,

> are adaptations not to scarcity but to abundance. They arise in populations that do not have significant material-scarcity problems with survival goods. We can observe gift cultures in action among aboriginal cultures living in ecozones with mild climates and abundant food. We can also observe them in certain strata of our own society, especially in show business and among the very wealthy.[25]

Abundance makes command relationships difficult to sustain and exchange relationships an almost pointless game. In gift cultures, social status is determined not by what you contol *but by what you give away*. 'Thus', Raymond continues, 'the Kwakiutl chieftain's potlach party. Thus the multi-millionaire's elaborate and usually public acts of philanthropy. And thus the hacker's long hours of effort to produce high-quality open source.'

Viewed in this way, it is quite clear that the society of Open Source hackers is in fact a gift culture. Within it, there is no serious shortage of the 'survival necessities' – disk space, network bandwidth, computing power. Software is freely shared. This abundance creates a situation in which the only available measure of competitive success is reputation among one's peers.

This analysis also explains why you do not become a hacker by calling yourself a hacker – you become one when *other* hackers call you a hacker. By doing so they are publicly acknowledging that you are somebody who has demonstrated (by contributing gifts) formidable technical ability and an understanding of how the reputa-

tion game works. This 'hacker' accolade is mostly based on aware-
ness and acculturation – which is why it can only be delivered by
those already well inside the culture. And why it is so highly prized
by those who have it.

Part III
Only connect . . .

Only connect! . . .
Only connect the prose and the
passion, and both will be exalted.

E. M. Forster, *Howards End*, 1910

14:
Web dreams

And yet, as angels in some brighter dreams
Call to the soul when man doth sleep,
So some strange thoughts transcend our wonted themes,
And into glory peep.

Henry Vaughan, 'Silex Scintillans', 1650

The Holy Grail of the computing business is something called the 'killer application' – a piece of software which is so self-evidently useful that it sells itself. The model of a Killer App is the first spreadsheet program, VisiCalc, written by Dan Bricklin, which provided a computerised way of doing the kind of cash-flow modelling routinely taught in every business school in America. It's the kind of tool everybody uses nowadays to put together a business plan or a departmental budget, but way back in 1977 all these things had to be done by hand, with a large piece of paper and a calculator.

Bricklin went to the Harvard Business School, where he was taught to do spreadsheet modelling with huge grids and lots of calculations. Whenever anything in one cell of the grid changed – which of course it always did when different scenarios were being considered – every other cell which depended on it had to be recalculated. It was an insane waste of time and effort, and Bricklin's great idea was that it should henceforth be done by computers.

In collaboration with a colleague called Bob Frankston, he wrote VisiCalc,[1] a program which essentially turned the computer screen into a window on a much larger virtual spreadsheet grid. But, in order to make the program into a marketable commodity, it had to be tailored to run on a particular computer, and this posed a problem because by that stage there were several – incompatible – machines on the market, notably the Apple II, Commodore's PET and Radio Shack's TRS-80 model. In the end, Bricklin and Frankston opted for Apple for the simple reason that someone had loaned them an Apple II.[2]

Nobody at Apple knew anything about this, naturally, so they were slightly perplexed by the kinds of people who suddenly began buying their machines. Like all early personal computers, the Apple II was aimed mainly at the hobbyist market; it sold initially to people who were interested in computing for its own sake rather than as a tool to enable them to do something else.

Many of the people who started buying Apple IIs after VisiCalc appeared, however, were not like this. For one thing, they wore suits. For another, they weren't much interested in what went on inside the machine. What suddenly began to happen was that executives would wander into computer stores across the US, see VisiCalc running, instinctively recognise its usefulness for budgeting and financial planning and say, 'I'll have one of those.' And when the salesman said that an Apple II was needed to run the program, they bought one of those too. After all, it was cheap enough to put on a consumables budget.

The first time this happened was a seminal moment in the history of computing – the instant when people realised that *it is software which sells hardware*, not the other way round.

On the Internet, the first Killer App was e-mail. And just as VisiCalc took Apple by surprise, electronic mail ambushed ARPA. For almost two decades, e-mail was the driving force behind the network's expansion. Left to itself, the Internet would probably have continued to grow at a healthy rate, but its expansion would have been limited by the fact that in order to use its facilities you had to master a fair amount of computerese. There was lots of wonderful stuff on servers all over the globe, but if you wanted to go beyond e-mail and get at it you needed to know about *FTP* and *GOPHER* and *binbex* and *directories* and *filenames* and a host of other user-hostile stuff.

All this changed at the end of the 1980s when a young Englishman named Tim Berners-Lee went back to work at CERN, the international particle-research laboratory in Geneva. His task was to find a way of helping physicists to use the Net more easily and effectively. He solved the problem and in the process invented a new way of structuring, storing and accessing information. He called it the World Wide Web.

Actually, it wasn't an entirely novel way of thinking about the problem. What Berners-Lee really did was to find a way of realising, on a global scale, an idea which had obsessed several other visionaries over half a century.

The first of these was Vannevar Bush, whom we met earlier when seeking the origins of the Net. During the Second World War, Bush was recruited initially to serve as President Roosevelt's Science Advisor and then as Director of the Office of Scientific Research and Development, in which role he had overall responsibility for *all* of the United States's scientific research during the war – including the development of microwave radar and the Manhattan Project which built the first atomic bomb.[3]

Bush's importance to the US war effort is almost impossible to overstate. One of Roosevelt's closest associates, Alfred Loomis, wrote that 'of the men whose death in the Summer of 1940 would have been the greatest calamity for America, the President is first, and Dr Bush would be the second'.[4]

In the 1930s Bush had done a lot of thinking and research not just on analog computing but also on the problems of storing and retrieving information. He was fascinated, for example, by the potential of microfilm – a technology which seems primitive and clumsy today, but which in the 1930s was revolutionary. MIT records[5] show that by 1937 he had constructed a prototype of a 'rapid selector' machine which could 'search reels of microfilm at speed and . . . copy selected frames on the fly, for printout and use'. This shows how far ahead of his time he was: most people were then preoccupied with using microfilm for storage; Bush was already thinking about how microfilm could be mechanically indexed and retrieved.

But something bigger was brewing in that fertile brain. The first hint of it came in an article he wrote in 1933 for the MIT journal, *Technology Review*. He writes of a machine with 'a thousand volumes located in a couple of cubic feet in a desk, so that by depressing a few keys one could have a given page instantly projected before him'.[6] There is also an extensive description of such a machine in a proposal Bush sent to Warren Weaver at the Rockefeller Foundation in 1937. And in 1939, shortly after he left MIT to head up the Carnegie Foundation, he wrote the first full draft of the article which was ultimately to trigger the Web.

Bush titled it 'Mechanization and the Record', and first thought of publishing it in *Fortune* magazine (and had some discussions with that organ's editorial staff about it), but then the war intervened and the manuscript remained in a drawer until 1945, when he dusted it off, gave it a new title and dispatched it to *Atlantic Monthly*, a magazine aimed at America's liberal intelligentsia. It was published in the July edition under the enigmatic title 'As We May Think'.

In Bush's papers in the Library of Congress there is a one-page memorandum in which he sets out the background to the article. 'My general objective', he wrote, 'was to influence in some small way thinking regarding science in the modern world, and to do so in an interesting manner. The secondary objective was to emphasize the opportunity for the application of science in a field which is largely neglected by organised science.'[7] His target audience was:

> the group of active thinkers who are taking part in affairs, rather than those who are developing a philosophy concerning the relation of science to human progress or the like . . . It seemed worth trying to influence the thinking of trustees of foundations, educational and scientific institutions, and those concerned with governmental support or control of scientific effort.[8]

Much of 'As We May Think' is taken up with envisioning ways of dealing with the information explosion which Bush saw as the concomitant of modern science and which had been an obsession of his since the 1930s. The general thrust of his argument was that machines would be needed to help mankind to cope with the explosion, and he expended a good deal of ink on discussing electro-mechanical devices which might help. These ideas may seem quaint to the modern reader – until one remembers that Bush was writing at a time when the digital computer was still a dimly perceived dream.[9] But strip away the antique 'how' in order to see *what* Bush was aiming at and you find a blueprint for what Tim Berners-Lee eventually implemented at CERN in 1989–91.

What concerned Bush was the inadequacies of paper-based indexing systems for retrieving information: 'The human mind does not work that way. It operates by association. With one item in its grasp, it snaps instantly to the next that is suggested by the association of thoughts, in accordance with some intricate web of trails carried by the cells of the brain.' He proposed a hypothetical device called the 'memex' – 'a sort of mechanised private file and library' in which an individual would store all his books, records and communications. It would be constructed to allow its contents to be consulted 'with exceeding speed and flexibility'. He envisaged that most of the memex's contents – books, newspapers, periodicals – would be purchased on microfilm ready for insertion. But he also envisaged a method of direct entry: 'On the top of the memex is a transparent platen. On this are placed longhand notes, photographs, memoranda, all sorts of things. When one is in place, the depression of a lever causes it to be photographed onto the next blank space in a section of the memex film, dry photography[10] being employed.'

Bush then goes on to describe various ways in which the memex's owner could scan the information it contained – there were levers for flipping forwards and backwards through the contents at speeds varying from one page to 100 pages at a time. There would be a 'special button' which would transfer the user to the first page of the index. Anything the user was looking at would be 'projected' on to a screen set into his desk.

So far, so conventional – though it's interesting to extract from Bush's dated talk of levers, buttons, projection and dry photography the essences of scroll bars, home pages, computer displays and scanning. What makes the article such a stunning read today, however, is his grasp of the essentials of associative linking. To Bush, the essential feature of the memex was a mechanism 'whereby any item may be caused at will to select immediately and automatically another'. This he called a 'trail'.

> When the user is building a trail, he names it, inserts the name in his code book, and taps it out on his keyboard. Before him are the two items to be joined, projected onto adjacent viewing positions. At the bottom of each there are a number of blank code spaces, and a pointer is set to indicate one of these on each item. The user taps a single key, and the items are permanently joined . . . Thereafter, at any time, when one of these items is in view, the other can be instantly recalled merely by tapping a button below the corresponding code space. Moreover, when numerous items have been thus joined together to form a trail, they can be reviewed in turn, rapidly or slowly, by deflecting a lever like that used for turning the pages of a book. It is exactly as though the physical items had been gathered together from widely separated sources and bound together to form a new book.

It is impossible to read this article, over half a century since it was written, and not feel that one is contemplating a blueprint for the Web. There is, in truth, little that is really new under the sun.

Bush's article created quite a stir, which is perhaps not surprising given that he was one of the most influential men in the United States. The Editor of the *Atlantic Monthly* penned a respectful preface to the article, likening it to Emerson's famous 1837 address on 'The American Scholar'. The Associated Press news agency put out an 800-word story about the piece on the day of publication. There was also a news story in *Time* magazine and a request from the Editors of *Life* to reprint a condensed version of the essay. In the eyes of his contemporaries, Bush was clearly no crackpot.

The *Life* edition of the piece appeared early in September 1945 with a subtitle: 'A Top U.S. Scientist Forsees a Possible Future World in Which Man-Made Machines Will Start to Think'. They also replaced the *Atlantic*'s numbered sections with headings and added specially commissioned illustrations.

By consenting to republication in *Life*, Bush was not only reaching the movers and shakers whose attention he craved, but a lot of other people besides. One of them was a young radar technician named Douglas C. Engelbart, who was stationed on a godforsaken island in the Philippines, awaiting a ship to take him home. 'I was a little navy boy,' Engelbart recalled many years later,

> an electronics technician in World War II out in the Philippines, and getting moved from one place to another. They stuck you on an island to wait to get you assigned to somebody else. Somebody said there was a library there. It was a Red Cross library up on stilts in a native hut, really neat, nobody there. I was poking around and found this article in *Life* magazine about his memex, and it just thrilled the hell out of me that people were thinking about something like that.[11]

Bush's ideas stuck in Engelbart's mind and remained there for five years, during which time he did research on wind tunnels for the agency which eventually became NASA. It was a job with great prospects at the time, but after five years he decided to give it up to pursue the dream which was to dominate the rest of his professional life – and which was to inspire successive generations of computer scientists. He decided, that instead of solving a particular problem, he would influence the very act of solving problems – that he would devote his life to a *crusade* (a word he often used later) to use computer power to augment human capabilities.

In pursuit of this astonishing dream, Engelbart quit his job, moved to the University of California at Berkeley and did a PhD in the new field of computer science. In 1957 he was hired by the Stanford Research Institute (SRI) – the lab which was the second node on the ARPANET. Two years later he got a small grant from the Air Force Office of Scientific Research, created a lab called the Augmentation Research Center and set out to change the world.

The part of Bush's vision which most grabbed Engelbart's attention was his fear that the accumulated knowledge of humanity would overwhelm our ability to handle it – that we would all be swept away by the 'information explosion'. Bush saw the threat clearly; but he could only fantasise about the technology that would be necessary to meet it. Engelbart, brooding on the same problem, had one advantage over his mentor: he realised that the development of the digital computer meant that the tools for 'augmenting' the human intellect were finally to hand.

Well, almost. This was in 1963, when computers cost hundreds of thousands of dollars and were programmed via stacks of punched cards by a white-coated priesthood. The remarkable thing about Engelbart was that he and his team not only envisioned but created the future that we now take for granted – bit-mapped screens, graphics-based interfaces, multiple windows,[12] software for creating documents structured as outlines, the 'groupware' needed for computer-mediated co-operative working, chorded keyboards and a host of other things. Oh – and he also invented the mouse, the ubiquitous pointing device which now roams a zillion desktops.

From our point of view, one of the most interesting features of the Augmentation system was the way it handled documents. As part of the Project Engelbart and his colleagues developed a system called NLS (for 'oN Line System') which enabled intricate linking from one document in an archive to others. This system was used to store all research papers, memos and reports in a shared workspace where they could be cross-referenced with each other.

Engelbart was that most elusive of creatures – a dreamer who got things done. He didn't just fantasise about his 'augmentation' technologies – in five years he constructed working systems which actually embodied them. There is a famous film,[13] for example, of him addressing a large audience at a Fall Joint Computer Conference in San Francisco in which he showed how a mouse and a special keypad could be used to manipulate structured documents and how people in different physical locations could work collaboratively on shared documents, online. It was, wrote one of those present, 'the mother of all demonstrations. As windows open and shut, and their contents reshuffled, the audience stared into the maw of cyberspace. Englebart, with a no-hands mike, talked them through, a calming voice from Mission Control as the truly final frontier whizzed before their eyes.'[14] The *coup de grâce* came when 'control of the system was passed, like some digital football, to the

Augmentation team at SRI, forty miles down the peninsula. Amazingly, nothing went wrong. Not only was the future explained, it was *there*, as Engelbart piloted through cyberspace at hyperspeed.'[15]

And the date of this demonstration? Why, 1968.[16] Bill Gates was twelve at the time; Steve Jobs was thirteen.

There are good grounds for saying that Engelbart is the patron saint of personal computing, and yet there is an inescapable poignancy about his career. Unlike many of the other pioneering hackers and engineers who created the computer industry he did not become rich (despite the twenty-odd patents he holds) and his fame extended only to those who understood the significance of what he had achieved. Most people today would recognise his name – if they recognised it at all – only in connection with his invention of the mouse.[17]

When Steven Levy was researching his book on the history of the Apple Macintosh, he went in search of Engelbart, and found him in a relatively obscure part of the Tymshare corporation (later purchased by McDonnell Douglas). Levy noted that the great man's office was a cubicle in one corner of a large room filled with file cabinets and similar warrens. 'The Moses of computers', he recalled, 'did not even rate an enclosed office.'

Engelbart is now the prime mover in an outfit called the Bootstrap Institute, housed at Stanford University. In his later years, he has been showered with distinctions by the industry he helped to create. He is no longer a prophet without honour in his own land, but there is nevertheless a disappointed edge to much of his contemporary musings.

Given what he set out to do, though, could it be otherwise? Engelbart's problem was that he wanted too much: he dreamed – as Wiener and Licklider dreamed – of making computer technology as powerful and as natural an extension of human capabilities as writing or talking. He asked the really hard question – *what* is it that we really want from this technology? – when most people are obsessed only with *how* to provide a particular kind of limited functionality.

The most difficult thing in life, said the great English management thinker Geoffrey Vickers, is *knowing what to want*. Engelbart, almost alone at the time, knew what he wanted from the technology. But to have achieved his goal would have required the kind of resourcing and multi-national drive which has pushed particle physics research to the frontiers it occupies today. Even ARPA – which funded and supported Engelbart for many years – did not possess that kind of clout. 'I confess that I am a dreamer,' he wrote in 1995. 'Someone once called me "just a dreamer". That offended me, the "just" part; being a real dreamer is hard work. It really gets hard when you start believing in your dreams.'[18]

The next intimation that the seed of the memex had germinated came in 1960 when a discombobulated genius called Theodor (Ted) Holm Nelson came up with the notion of 'nonsequential writing' – an idea for which, five years later, he coined the term *hypertext* at a conference of the Association of Computer Machinery.

Hypertext is an implementation of the idea that texts are effectively nonlinear, which is itself a recognition of the fact that when reading documents we often jump to other locations. For example, one might go to a footnote, look up a definition, switch to another section (either because it's mentioned or because we're no longer interested in the current section), or switch to another document (because it's mentioned, because an idea in one document reminds us of another document . . .).

In essence therefore, the idea of hypertext is not a new one. It could be argued, for example, that the Talmud – a Biblical commentary (and meta-commentary), in which the authors refer not just to passages of the bible, but also to other commentaries – is effectively a hypertext, as are many other ancient documents. One could make much the same claim for James Joyce's *Ulysses* – as witnessed by the innumerable commentaries and editions containing detailed exegeses of the references implicit in the text.[19]

Ted Nelson is a legendary figure in the computing industry because of his single-minded pursuit of a dream. His life has revolved around the Xanadu project – an attempt to construct a global hypertext publishing system – once described by Gary Wolf as 'the longest-running vaporware[20] story in the history of the computer industry'.

> It has been in development for more than 30 years. This long gestation period may not put it in the same category as the Great Wall of China, which was under construction for most of the 16th century and still failed to foil invaders, but, given the relative youth of commercial computing, Xanadu has set a record of futility that will be difficult for other companies to surpass. The fact that Nelson has had only since about 1960 to build his reputation as the king of unsuccessful software development makes Xanadu interesting for another reason: the project's failure (or, viewed more optimistically, its long-delayed success) coincides almost exactly with the birth of hacker culture. Xanadu's manic and highly publicised swerves from triumph to bankruptcy show a side of hackerdom that is as important, perhaps, as tales of billion-dollar companies born in garages.[21]

Nelson grew up in New York, the only – and lonely – child of an absentee, movie-directing father and an actress mother. A dreamy, unathletic boy, he was raised by his elderly grandparents, who must have been baffled by their obviously brilliant but undisciplined charge. His hatred of conventional structure made him difficult to educate. It is said that once, bored and disgusted by school, he plotted to stab his seventh-grade teacher with a sharpened screwdriver, but lost his nerve at the last minute and instead walked out of the classroom, never to return. On the long walk home, he came up with the four maxims that have guided his life: 'most people are fools, most authority is malignant, God does not exist, and everything is wrong'.[22]

Not surprisingly, Nelson's heroes tended to be non-conformists – people like Buckminster Fuller, Bertrand Russell, Walt Disney, H. L. Mencken and Orson Welles. He went to Harvard and studied strategy as a graduate student under Thomas Schelling but found himself increasingly obsessed by the way traditional writing and editing imposed a false and restrictive order on textual material. Nelson had little interest in the progressive narratives embodied in books; he wanted everything to be preserved in all its chaotic flux, ready to be reconstructed when needed.

At some point along this voyage of discovery, Nelson discovered 'As We May Think' and made the vital connection: the technology needed to drive the associative linking envisioned by Bush had been invented. There was no need for levers and microfilm and projectors. All that was required was a digital computer.

Nelson's passion was fuelled by his unconventional, almost tormented, make-up. The lonely child, raised in an unconventional family, was father to the man who became:

> a rebel against forgetting, and a denier of all forms of loss and grief. (Some of Nelson's disciples would one day take this war against loss even further, and devote themselves to the development of cryonic technology for the freezing and preservation of corpses.) Tormented by his own faulty memory, Nelson developed the habit of asserting that only a technology for the preservation of all knowledge could prevent the destruction of life on Earth. The thought that some mental connection or relationship might dissolve was unbearable. Not only was the constant churn and dispersal of his own thoughts personally devastating, but the general human failure to remember was, Nelson thought, suicidal on a global scale, for it condemned humanity to an irrational repetition of its mistakes.[23]

The first working hypertext system was developed at Brown University in 1967 by a team led by Andries van Dam. The Hypertext Editing System ran in 128K memory on an IBM/360 mainframe and was funded by IBM, who later sold it to the Houston Manned Spacecraft Center, where it was used to produce documentation for the Apollo space program. In 1968 van Dam developed the system further into a multi-user time-sharing version, still running on the IBM mainframe which was later commercially implemented by Philips.

In 1981 Nelson decided to expand the idea into *Xanadu* – a central, pay-per-document hypertext database encompassing all written information. He got the name from Coleridge's 'Kubla Khan', in which Xanadu is a 'magic place of literary memory'. Nelson's dream was to create a unified literary environment on a global scale, a repository for everything that anybody has ever written. He called it a 'docuverse' (document universe) where 'everything should be available to everyone. Any user should be able to follow origins and links of material across boundaries of documents, servers, networks, and individual implementations. There should be a unified environment available to everyone providing access to this whole space.'[24]

One of the strangest features of the Xanadu system is that it has no concept of deletion. Once something is published, it is for the entire world to see for ever. As links are created by users, the original document remains the same except for the fact that a newer version is created which would have references to the original version(s). But it turns out that conventional computer file systems are unable to implement such a scheme and so Xanadu programmers were compelled to focus much of their effort on the redesign and reimplementation of file systems – which in turn has required the creation of a whole new operating system incorporating a hypertext engine. In the circumstances the surprising thing about Xanadu is not that it is so far behind schedule but that it is still going at all.

The strange thing about dreams is that they are contagious. Programmers are particularly vulnerable to infection because the stuff with which they work is so magically fluid. Engineers in other fields are caged by the laws of physics: there is no way that water can be persuaded to flow uphill, nor heat energy to transfer from a cold body to a hotter one. There is no such thing as a frictionless bearing or acceleration without force; no rocket can 'slip the surly bonds of earth' without liberating the chemical energy pent up in its fuel tanks.

But in software – ah! that's different. In software you can program a set of pixels[25] to move endlessly in a circle, and they will continue to do that for ever if you leave them to it. They need neither fuel nor food, and they will never complain. In that sense, being a programmer is like being Napoleon before the retreat from Moscow. Software is the only medium in which the limits are exclusively those set by your imagination. And even the fact that computing hardware is not powerful enough to deliver what our program needs doesn't matter, because one day, given Moore's Law (the one which states that computing power doubles every eighteen months to two years), it will; and even if it doesn't, you can always run your software on a 'virtual machine' – that is, a computer simulation of an actual machine. You can do anything with a program, implement any dream, given only one thing – that you are smart enough.

Silicon Valley, that expensive patch of real estate centred on the town of Palo Alto, thirty miles south of San Francisco, has more dreamers per square yard than any other place on earth, which is partly because it is stuffed full of software wizards. Chief among them in the mid-1980s was a programmer called Bill Atkinson who was already a legend in the business as the author of MacPaint, the graphics program which came bundled with early versions of the Apple Macintosh.

In an age of Windows and Macintosh computers, we have become blasé about drawing and painting packages. But many of us will never forget our first encounter with Atkinson's baby. In my own case, it happened in 1984 at a workshop for academics known to be interested in personal computing which was organised by Apple UK at the University Arms hotel in Cambridge.

The venue was a stuffy conference suite ringed with tables covered in green baize. On each table stood a chic little machine with a nine-inch screen and a detached keyboard. Compared with the clunky, three-box design which then represented the industry's idea of what a personal computer should look like, these elegant little machines seemed, well, just *gorgeous*. I couldn't wait to get my hands on one.

After an initial spiel by the Apple crowd, we were let loose on the machines. They had been set up, for some unfathomable reason, displaying a picture of a fish. It was, in fact, a MacPaint file. I remember staring at the image, marvelling at the way the scales and fins seemed as clear as if they had been etched on the screen. After a time I picked up courage, clicked on the 'lassoo' tool and selected a fin with it. The lassoo suddenly began to shimmer. I held down the mouse button and moved the rodent gently. *The fin began to move across the screen!*

Then I pulled down the Edit menu, and selected Cut. The fin

disappeared. Finally I closed the file, confirmed the decision in the dialog box, and reloaded the fish from disk. As the image reappeared I experienced an epiphany: I remember thinking, *this is the way it has to be.* I felt what Douglas Adams later described as 'that kind of roaring, tingling, floating sensation' which characterised his own first experience of MacPaint.[26] In the blink of an eye – the time it took to retrieve the fish from disk – all the teletypes and dumb terminals and character-based displays which had been essential parts of my computing experience were consigned to the scrapyard. I had suddenly seen the point – and the potential – of computer graphics.

All this was Bill Atkinson's doing. In the circumstances, to call him a programmer is like calling Christian Dior a dressmaker. Atkinson is in fact a genius, or at the very least an artist whose medium just happens to be software. Indeed, this is sometimes how he describes himself. 'The art of creating software that is usable by individuals', he once said,

> is a communication skill. It is not a programming skill. Programming and what a software artist does is analogous to a pianist who needs to know how to move the keys and have that down cold so that he can concentrate on the feeling and message that he is portraying in his rendition of this music. So slinging the bits is an enabling technology for me to express and communicate and teach. The most exciting thing for me is when I see people amazed and pleased at the newfound power they got from a program – where they say 'Wow, I can do this!' That's the feeling people got back in 1984 when they saw MacPaint and started using it.[27]

There is another respect in which the artistic stereotype fits Atkinson. Whereas many of his contemporaries talk about the *market* for their products, Atkinson talks about the *audience* for his. The most important, most exhilarating thing about MacPaint from his point of view was that it was (initially) bundled with *every* Mackintosh sold. And, like many artists, he is acutely sensitive about getting proper recognition for his work. One of the reasons he left academic life in 1978, for example, was because he felt that the publications emanating from the project on which he worked did not give him due credit for his contributions. Later, when he worked on the Lisa (the precursor to the Macintosh), Apple management chose to highlight the role of the Lisa project managers rather than that of the hackers who created its software. Atkinson was so disgusted that he defected and joined the small team of desperadoes who were designing the Macintosh.

After the launch of the Mac and the triumph of MacPaint, Atkinson used his new-found freedom as an Apple Fellow to embark on the most ambitious project of his life, and in so doing joined the elite club of which Bush, Wiener, Licklider, Engelbart and Nelson were founder members – an association of those trying to realise the dream of harnessing computing power to improve the human condition.

Atkinson's name for his dream was 'Magic Slate'. In part, it was derived from Alan Kay's concept of the 'Dynabook'[28] – a computer so useful and easy to use that its owner could not bear to be parted from it. It would be no bigger than an A4 (paper-based) notebook, have a high-resolution screen, accept handwritten input and be so cheap that a user could afford to lose one without being unduly impoverished.

Magic Slate was what Atkinson thought Apple should do next and he was desperately upset to discover that the company's new management under John Sculley – a former Pepsi-Cola executive– was no longer interested in the realisation of dreams. Atkinson withdrew to his house in deep depression and stayed there for several months, unable to work, too dejected even to turn on his computer. Then, according to Steven Levy,

One night, he stumbled out of his house in the Los Gatos hills, wandering aimlessly. The night was cloudless, gorgeous. He spent hours gazing into the sky. Suddenly his own troubles fell into perspective. Who was he? A stray pixel in the eternal bit map. There was no way that human beings were alone in this universe, he thought, and no way that we're the pinnacle of all possible life-forms. The thought was oddly encouraging. Given this fact, it seemed to him, the important thing was what one *could* do. Could you aid your fellow humans, enhance their abilities? Make a contribution?[29]

Atkinson gave up on Magic Slate, but translated some of the ideas behind it into software which could be delivered on a real machine – the Mac. He exchanged the page metaphor of Slate for that of the oldest information-handling tool of all – the 3×5-inch index card found in every library in the world. Atkinson realised that if you added computing power to these cards they could be transformed into something quite magical. Given the right software, a user could instantly install a link from one card to another. A series of such links could set up an information 'trail' of the kind envisaged by Vannevar Bush. Thus was born the concept which eventually became HyperCard.

According to Atkinson, the ideas behind HyperCard sprang from several sources.[30] One was a Rolodex program he had written some years earlier to keep track of his own references. Another was some work he had done on compression algorithms, while a third source was his experiments with more efficient searching algorithms. 'In my research,' he recalled,

> I had already figured out that, at least theoretically, the searching could be speeded up 100 times. When I actually got to doing it, the measured performance was 700 times faster! This breakthrough allowed us to search the Los Gatos town library card catalog – which had 100,000 cards or 15 megabytes of text – in 2 seconds instead of 10 minutes. We were really pleased. It was very exciting when that first broke.[31]

The core of HyperCard was the card. Each card could have user-installed 'hot spots' which, if clicked upon, would instantly cause a jump to another card. In one of the demonstration stacks, for example, clicking on the image of a hat took one to a set of cards about hats.

A collection of linked cards was called a 'stack'. The key card in the stack was called the 'Home' card. And while links which involved just simple branching could be set up by pointing and clicking, discussions with an Apple colleague, Dan Winkler, led to the idea of integrating user-definable functions and building in a programming language (called HyperTalk) which would make HyperCard a whole software environment enabling the construction of quite complex programs.

With Alan Kay's help, Atkinson sold the idea of HyperCard to John Sculley. A key factor in persuading Atkinson to overcome his suspicion of Apple's management following the rejection of Magic Slate seems to have been Sculley's agreement to bundle HyperCard with every Mac – just as had been done with MacPaint in the early days. Using his home as a programming lab, Atkinson and a team of several hackers created a robust implementation of HyperCard which Sculley himself launched at the August 1987 Macworld Expo.

HyperCard was, as Levy maintains,

> a brilliant exploitation of the Macintosh's abilities. The program made it abundantly clear that the Mac was the first engine that could vivify the hypertext dream – one could hack out Vannevar Bush-like information trails by a few simple manipulations and clicks of the mouse . . . It seemed in some respects like the dawning of a new era, the age of hypermedia, where the common man would not only have tremendous access to previously elusive shards of information, but would actually become a master manipulator of that information.[32]

Atkinson himself saw HyperCard as a logical extension of the Macintosh mission – something which would give power to users that had previously belonged only to professional programmers. The Macintosh dream was about putting the power of the personal computer into an individual person's hands without requiring them to become computer experts. But, paradoxically, in order to do that its designers had created a relatively *closed* machine in both hardware and software terms compared to the IBM PC. In order to write applications for the Macintosh, for example, programmers had to be intimately familiar with the Toolkit which Atkinson and his colleagues had crafted and embedded in a chip inside the machine. To make the most of Macintosh's features you had to spend months studying the *Inside Macintosh* handbook to understand how to use all the Toolkit features, the graphics, menus and so on. The result, thought Atkinson, was the dilution of the Macintosh dream, because the individuals couldn't get all the power of the personal computer. They could only use 'canned pieces of power'. But HyperCard,

> acting like a software erector set,[33] really opens up Macintosh software architecture to where individual people can make their own customised information environment, and interactive information and applications without having to know any programming language. It takes the creation of software down to the level of MacPaint images that you like, then pasting buttons on top of them to make them do what you want. HyperCard puts this power into the hands of any Macintosh user.[34]

In the two decades since personal computers appeared, there have only been a few pieces of seminal software. What is Word 97 but a better Wordstar? What is Excel but a smarter VisiCalc? Modern database programs are mostly just dBASE II with bells and whistles.

In these terms, HyperCard was a truly fresh idea: it was completely original, and it seemed to offer a new way of using computers for important and difficult tasks. It was – and is – a wonderful piece of software which gives people who know nothing about programming the 'freedom to associate' – to organise information in ways which match or suit their own cognitive styles.

It is also a terrific 'erector set' (to use Atkinson's phrase) for mocking up ideas about interfaces or presentations or databases. But, although it has had a great impact on those of us who think about such things, it has not had the seismic impact its creator expected. 'It's the largest thing I've attempted,' he once said, 'and I think it's the most significant in terms of what it will do to the computing community as a whole . . . HyperCard is going to open up the whole meaning of what personal computers can be.'[35]

The sad truth is that it didn't – and for one very simple reason. It was rooted in the notion that the computer was a standalone device, complete in itself. HyperCard enabled one to obey E. M. Forster's injunction – *only connect* – but it assumed that all the connections worth making resided on your hard disk. Bill Atkinson could not have known it, but something was brewing which was to render that assumption absurd.

15:
Liftoff

I regularly surf the Internet. It is by far the most important innovation in the media in my lifetime. It's like having a huge encyclopedia permanently available. There's a tremendous amount of rubbish on the world wide web, but retrieval of what you want is so rapid that it doesn't really matter. This, of course, isn't true of a newspaper where you have to wade through the rubbish in order to find the good bits.

Richard Dawkins[1]

Of all the places in the world which might have generated a Killer App, CERN is surely the least likely. The European Centre for Nuclear Research is a giant organisation devoted to research into the fundamental sub-atomic particles which make up matter. It runs a vast underground torus seventeen miles in diameter under the city of Geneva in which particles are accelerated almost to light-speed and then smashed into other particles. The patterns resulting from these collisions tell high-energy physicists something about the nature of physical reality.

As a multi-national organisation, CERN is necessarily bureaucratic and very grown-up. It has to be because it runs up huge bills which are paid by men in suits sent from the governments of the participating states. It was the polar opposite of the anarchic Palo Alto start-ups which spawned the personal computer business. (A standing joke at the time went: what is the difference between

Apple Corporation and the Boy Scouts? Answer: the Scouts have adult supervision.)

CERN had lots of adult supervision. And yet it was here in a few magical months between March 1989 and November 1990 that the World Wide Web was spun.

Its founding spider was Tim Berners-Lee. He is not, on the face of it, a charismatic figure. *Au contraire*: he is a youthful fortysomething who dresses neatly but casually, drives a Volkswagen and has none of the vices traditionally associated with great men. Yet an aura clings to him, generated not just by what he has achieved, but even more so by what he has chosen not to do. For this is a man who invented the future, who created something that will one day be bigger than all the other industries on earth. This is a man whose intellectual property rights could have made him richer than Croesus.

And yet he turned his back on all that to work for the common good. Berners-Lee now operates from a spartan office at MIT's Computer Science Lab, where he draws a modest salary as head of the World Wide Web Consortium (W3C), the body which tries to maintain technical order in the burgeoning technology of the Web.[2] No wonder that, whenever he appears in public, the images that come to mind are not those from Kubrick's film *2001*, but those from 'The Road Not Taken', Robert Frost's wonderful poem about the choices we make in life.

> Two roads diverged in a wood. And I –
> I took the one less travelled by,
> And that has made all the difference.

Berners-Lee has software in his blood. Both his parents were programmers who worked for the British company Ferranti on one of the first commercial computers. He read physics at Oxford, where he built his first computer with a soldering iron, an M6800 microprocessor chip and an old television set. Graduating in 1976, he worked first for Plessey and later for a firm writing typesetting software.

In many respects, he looks like an Englishman from central casting – quiet, courteous, reserved. Ask him about his family life (he has an American wife and two children) and you hit a polite but exceedingly blank wall. Ask him about the Web, however, and he is suddenly transformed into an Italian – words tumble out nineteen to the dozen and he gesticulates like mad. There's a deep, deep passion here. And why not? It is, after all, his baby.

The strange thing is that it all happened because he has a lousy memory. Names and faces often elude him, for example. 'I needed something to organise myself,' he says, 'I needed to be able to keep track of things, and nothing out there – none of the computer programs that you could get, the spreadsheets and the databases, would really let you make this random association between absolutely anything and absolutely anything.'

So in the end he wrote such a program himself, while on a six-month consultancy at CERN in 1980. He called the program ENQUIRE (for 'enquire within about everything'). Berners-Lee describes it as a 'memory substitute' which enabled him to fill a document with words which, when highlighted, would lead to other documents for elaboration.

> It allowed one to store snippets of information, and to link related pieces together in any way. To find information, one progressed via the links from one sheet to another, rather like in the old computer game 'adventure'. I used this for my personal record of people and modules. It was similar to the application *HyperCard* produced more recently for the Apple Macintosh.[3]

When his attachment ended, Berners-Lee returned to England and helped found a start-up company doing computer graphics and related stuff. But the lure of CERN endured and he returned there in the late 1980s charged with supporting the lab's community of physicists in their retrieval and handling of information. What struck him second time around was how, at an institutional level, the laboratory suffered from the same memory problems as himself.

CERN is a vast organisation, doing research of unimaginable complexity. Much of its experimentation is done by teams of visiting physicists who come and go. Maintaining coherent documentation in such circumstances was a nightmarish task. Berners-Lee felt that an organisation like CERN – with so many people coming and going and leaving little trace of what they'd done and why they'd done it – needed some space in which it could consolidate its organisational knowledge.

The task he set himself was to convince the management that a radically new approach to the structuring of information was required to meet the needs of such an idiosyncratic organisation. So at the beginning of 1989 he sat down at a Macintosh and hammered out a proposal that would change the world. He began by pointing out that many discussions of the future at CERN ended with the question: 'Yes, but how will we ever keep track of such a large project?' He claimed to have an answer to the question. But before revealing it he wanted to explore the problems that people at CERN experienced in keeping control of the information they possessed. The central difficulty, he argued, was that information was continually getting lost.

> The actual observed working structure of the organisation is a multiply connected 'web' whose interconnections evolve with time. In this environment, a new person arriving, or someone taking on a new task, is normally given a few hints as to who would be useful people to talk to. Information about what facilities exist and how to find out about them travels in the corridor gossip and occasional newsletters, and the details about what is required to be done spread in a similar way. All things considered, the result is remarkably successful, despite occasional misunderstandings and duplicated effort.[4]

CERN, by its very nature, had a high turnover of people because physicists from participating countries were continually coming and going, staying on average only two years. The introduction of the new people demanded a fair amount of their time and that of others before they had any idea of what went on. The technical details of past projects were sometimes lost for ever, or only recovered after a detective investigation in an emergency. Often, the information *had* been recorded but simply could not be located.

If a CERN experiment were a static, once-and-for-all event, all the relevant information could conceivably be contained in one enormous reference book. 'But', he observed, as it is, CERN is constantly changing as new ideas are produced, as new technology becomes available, and in order to get around unforeseen technical problems. When a change is necessary, it normally affects only a small part of the organisation. A local reason arises for changing a part of the experiment or detector. At this point, one has to dig around to find out what other parts and people will be affected. Keeping a book up to date becomes impractical, and the structure of the book needs to be constantly revised.[5]

Examples of the kinds of information needed were: 'Where is this module used? Who wrote this code? Where does he work? What documents exist about that concept? Which laboratories are included in that project? Which systems depend on this device? What documents refer to this one?'[6] What was needed, he argued, was some kind of linked information system.

But what sort of system? After criticising the types of linking which were then in vogue – for example, the hierarchical tree-structures exemplified by the 'help' systems of minicomputers; or those which relied upon indexed keywords – Berners-Lee went on to propose a solution to CERN's problem. It was hypertext.

The special requirements of a hypertext system for CERN were, he believed, that it should: allow remote access across networks; be heterogeneous (that is, allow access to the same information from different types of computer system); be non-centralised; allow access to existing data; enable users to add their own private links to and from public information, and to annotate links as well as nodes privately; and enable 'live' links to be made between dynamically changing data.

Finally, Berners-Lee proposed three specific applications areas within CERN for such a system: development project documentation; document retrieval; and personal skills inventories (important in an outfit with a large floating population of differently skilled experts).

The paper concludes with a recommendation that CERN should 'work toward a universal linked information system in which generality and portability are more important than fancy graphics techniques and complex extra facilities'.

The aim should be to allow a place to be found for any information or reference which one felt was important, and a way of finding it afterwards. The result should be sufficiently attractive to use that the information contained would grow past a critical threshold, so that the usefulness of the scheme would in turn encourage its increased use. The passing of this threshold [should be] accelerated by allowing large existing databases to be linked together and with new ones.[7]

To test these ideas he proposed a practical project to build a prototype system along these lines. 'I imagine', he wrote,

that two people for 6 to 12 months would be sufficient for this phase of the project. A second phase would almost certainly involve some programming in order to set up a real system at CERN on many machines. An important part of this, discussed below, is the integration of a hypertext system with existing data, so as to provide a universal system, and to achieve critical usefulness at an early stage.[8]

And yes, he added, in a parting aside, 'this would provide an excellent project with which to try our new object oriented programming techniques!' – which his fellow hackers immediately recognised as a reference to his fancy new NeXT computer.

Berners-Lee's proposal was first circulated in March 1989 with a request for comments. In May the following year, it was recirculated, presumably with amendments. In October 1990 the project proposal was reformulated with encouragement from CERN senior management and Robert Cailliau as co-author. It was at this point that the central difference between Berners-Lee's ideas and those of Engelbart, Nelson and Atkinson became clear, for the name 'World Wide Web' was chosen, implying that the Net was an integral part of the concept from the word go. (The use of the term 'web' was not original because it had already been long established in the hypertext community; for example, the InterMedia system developed at Brown University in the early 1980s had a 'Web tool' in its toolbox.)[9]

Thereafter, the pace of events speeded up. In November 1990, Berners-Lee sat down at his NeXT workstation and hacked out a program called a browser (see below) which provided a virtual 'window' through which the user saw the 'web' of linked resources on the Internet. It refracted, as it were, a world of disparate information sources in such a way that they appeared as a uniform whole.

Since the NeXT machine was pretty exotic hardware and Berners-Lee had set as one of his design criteria that the Web should work over a range of different computer systems, there was clearly a need for another kind of browser which would function on non-graphical displays. Accordingly a newly arrived technical student from Leicester Polytechnic called Nicola Pellow was set to work writing a simple 'line-mode' browser.

By Christmas 1990 demo versions of both browsers and a prototype Web server were available, enabling users to access hypertext files, articles from Internet News Groups and files from the help system of one of the CERN computers. By March of the following year, the line-mode browser was released to a limited audience to run on the more powerful workstations in the CERN environment. On 17 May 1991 the WWW software was generally released on central CERN machines. In August, information about the project and the software was posted in relevant Internet news groups like alt.hypertext, comp.sys.next, comp.text.sgml and comp.mail.multi-media. In October gateways were installed enabling the browsers to access Help files on the laboratory's Digital VAX computers and the 'Wide Area Information Servers' (WAIS) machines which were then the nearest thing to search engines that the Internet possessed. Finally in December, the CERN computer newsletter announced the Web to the world of High Energy Physics. It had taken just over a year from the moment Berners-Lee had typed the first line of code.

Berners-Lee's underlying model of the Web was what is known in computer-speak as a 'client–server' one. That is to say, he envisaged a system in which information would be held on networked computers called *servers*, and that these would be accessed by *client* programs (browsers) running on other net-worked computers. Servers, in this model, are essentially givers, while clients are always takers (though they give some information about themselves to servers at the moment of interaction). The central tasks in building such a system were to write the programs which would enable computers to act as servers and clients, to create a common language in which both could converse and to set up some conventions by which they could locate one another.

The client–server model was already well established in the computer business when Berners-Lee started work. There were innumerable machines on the Net which operated as servers, and there were several ways of extracting information from them. At the lowest level, you could use the primitive TELNET facility to log on to a remote machine and (if you had the necessary permissions) run programs on it. Or you could use the File Transfer Protocol (FTP) program to log on remotely and download files.[10] And there were various search facilities[11] – the precursors of search engines – for locating information on the Internet.

In order to make use of these facilities, however, you needed to know what you were doing. Accessing the Net before Berners-Lee was akin to using MS-DOS or UNIX – you could do almost anything provided you knew the lingo. The trouble was that the lingo was user-hostile. Computer freaks took to it like ducks to water; the rest of humanity, however, looked the other way. The pre-WWW Net was, wrote Robert Reid,

an almost militantly egalitarian and cooperative community. No-body owned the network. Virtually nobody made money from it directly. Almost every piece of software that governed or accessed it was free (the people who wrote it generally did so from the goodness of their hearts, or to make names for themselves, or as parts of funded projects). But its egalitarianism aside, the Internet's tight de facto admission requirements of technical acumen, access and pricey tools also made it a very elite realm.[12]

One of the central tasks Berners-Lee faced in creating the Web was the lowering of this threshold. He achieved it partly by inventing an *interface* – a program which stood between the user and the vast and disparate information resources of the Net.

In seeking a model for this interface he drew heavily on ideas which had emanated from the hypertext community, where the problem of navigating through a virtual space of linked texts had been addressed through the notion of a 'browser' – that is, a virtual window which displayed the structure of the space. The Xerox *NoteCards* system developed in the early 1980s, for example, had a special 'browser card' which displayed editable node-link diagrams showing the structure of some portion of the hypertext network currently in use.[13] Other hypertext or hypermedia systems had similar tools.

But to the original hypertext notion of browsing Berners-Lee added something else which echoed the thinking of Ted Nelson, namely the idea that *documents should be editable by their readers*. Berners-Lee's idea of a browser therefore was not just an interface that would provide passive viewing, but something that would allow users to create their own inline links even as they were reading a document. This is the one part of his vision which did not survive beyond the CERN version (though it lives on to some extent in the AOLPress Web-editing program).[14]

Because NeXT workstations (and indeed graphics terminals) were then still a rarity in the networked computing world, it was essential that a non-graphical browser also be developed. This is what Nicola Pellow had been charged with creating – a program which would run on character-based screens, presenting Web resources as a set of items in a menu. Such a program had several advantages, not least its relatively compact code size (which made it easy to run and download) and the fact that it made few demands on the machines which would run it.

Creating the Web was not just a matter of writing the code for browsers, however. Because the Net, with its incredible diversity, was central to the project, Berners-Lee had to invent a way of ensuring that publicly available information resources held on any networked computer anywhere in the world could be accessed through the browser. The only way to do this was to create a set of *protocols* by which different machines could talk to one another and exchange information. One protocol (analogous to the IP convention which specifies the unique address of every machine connected to the Net) had to specify the *location* at which information was held. For this Berners-Lee invented the *Uniform Resource Locator* or URL.

Another protocol was needed to specify how information-exchange between machines should be handled. For this he created the *Hypertext Transport Protocol* (HTTP), which was analogous to FTP. And finally he had to invent a uniform way of structuring documents. For this he proposed *Hypertext Mark-up Language* or HTML as a subset of the Standard Generalised Mark-up Language (SGML) tagging system which was already established in the electronic publishing business.

For a language which has become the lingua franca of the wired world, HTML is remarkably simple – essentially consisting of a set of conventions for attaching tags to text. A Web page is just a text document in which certain elements of the text are tagged in particular ways. If I wanted to make this sentence appear boldfaced in a browser window, for example, I would type the following on the Web page:

> < B > If I wanted to make this sentence appear boldfaced in a browser
> window, for example, I would type the following on the Web page
> < /B >

In HTML-speak, < B > means 'start displaying boldface here'; < /B > cancels that injunction and returns to displaying normal type. If I want the word 'here' to serve as a link to another Web page (perhaps held on a server on the other side of the world), I tag it like this:

> Click < A HREF="http://www.kpix.com/live/" > here < /A > to see a
> live panoramic photograph of San Francisco < /A >

'HREF' is the HTML code for 'hyperlink reference'; 'http://' indicates that this is a Web address; and 'www.kpix.com/live/' is the URL to which the reference points. The first part of the URL – 'www.kpix.com' – corresponds to the Internet address of the server containing the desired images, while the '/live/' portion refers to a particular area (for example, a directory) on the server's hard disk.

URLs and HTML are thus pretty straightforward. HTTP – the protocol which controls how machines issue and respond to requests for information – is more complicated.[15] In non-technical language, what HTTP essentially does is to prescribe how the four stages of a Web transaction – connection, request, response and close – should be conducted.

Looking back, it is not so much the elegance of Berners-Lee's creation which makes one gasp as its blinding comprehensiveness. In just over a year he took the Web all the way – from the original conception, through the hacking out of primitive browsers and servers, to the creation and elaboration of the protocols needed to make the whole thing work. And on the seventh day he rested.

The Web went public on 15 January 1991 when the line-mode browser developed at CERN was made available by the process known as 'anonymous FTP'. That is to say, anyone with a Net connection and a copy of the FTP program could call up the CERN site, log in without having to use a password and download the browser code. Other researchers – many not based at CERN – were already busy developing graphics-based browsers which did not require fancy workstations. In April a Finnish browser for UNIX called 'Erwise' was released and this was followed in May by the release of Pei Wei's Viola graphical browser (also for UNIX machines). By July, CERN was distributing all the code (server and browsers, including Viola) from its servers.

In November 1992 there were twenty-six known WWW servers in existence, including one at the US National Center for Supercomputer Applications (NCSA) at the University of Illinois at Champaign-Urbana. Two months later, the number of servers had almost doubled – to fifty. By Internet standards this was encouraging growth, but nothing spectacular: the Web was still just one of the many applications running across the network. Compared to e-mail, FTP and all the other stuff running across the Net, the HTTP traffic was still pretty small beer. And then, as Yeats might have said, all was changed, changed utterly. In the spring of 1993 a student in Illinois launched a browser which turned the Web into the Killer App of all time. It was called Mosaic.

A salutary thought, one which ought to be engraved in 96-point Helvetica Bold over the desk of every tenured academic, is that many of the most important things in the evolution of the Net – for example, some of its original protocols and the consultation process through which they were refined – were invented not by professors, but by students.

Mosaic was no exception. Its guiding spirit, ringmaster and general-purpose slave-driver was a six-foot-four undergraduate called Marc Andreessen who looks like a kinsman of Garrison Keillor and has, according to those who know him, a gargantuan appetite for pizza, Bach, newsprint, algorithms, ideas, John Barth, Vladimir Nabokov, images, UNIX code and communications bandwidth.[16] Andreessen was born and brought up in New Lisbon, a small town in Wisconsin, the child of a seed salesman father and a mother who worked as a Land's End shipping clerk. He showed early promise – teaching himself BASIC from a library book at the age of nine, for example. He attended a small high school and then moved on to the University of Illinois, where he decided to study computer science because 'electrical engineering was too much work'.[17]

Like many US students from modest backgrounds, Andreessen needed to work his way through college, so he took a job paying $6.85 per hour at the National Center for Supercomputer Applications. As its name implies, NCSA specialised in heavy-duty computing. The federal government had decided years earlier that supercomputers were so expensive that even the United States could afford only a few of them. Accordingly a number of centres were set up and equipped with powerful number-crunchers made by Thinking Machines, Cray and other corporations specialising in computational horsepower. The brief of each centre was to tend its machines and make their facilities available via the Net to federally funded researchers across the US.

The NCSA was the grandaddy of these supercomputing centres, a veritable hacker's paradise. And yet Andreessen was, frankly, underwhelmed by the place. 'When I was there,' he later recalled, 'it had been around for roughly eight years or so and, at that point, it had a very large established budget – many millions of dollars a year – and a fairly large staff and, frankly, not enough to do.'[18]

What had happened, of course, was that Moore's Law had begun to catch up with supercomputers; the gap between their computational power and that of the advanced workstations being rolled out by Sun, Hewlett-Packard, Silicon Graphics and Digital had narrowed dramatically. But the budgetary momentum of large federal programmes kept the Supercomputing Centres going long after their Unique Selling Proposition had evaporated. .

The main driving force behind the creation of Mosaic was probably boredom. Andreessen got fed up writing 3-D visualisation code on a Silicon Graphics Indy machine and started to look at what else was available. The strange thing was that, in this Mecca of techno-freaks, the thing that most impressed him was the Internet. And yet even that suffered from what he perceived as a glaring flaw: for all its potential, the access software needed to use the Net was at least ten years behind the mainstream computer industry. 'PC *Windows* had penetrated all the desktops,' Andreessen told an interviewer,

> the Mac was a huge success, and point-and-click interfaces had become part of everyday life. But to use the Net you still had to understand Unix. You had to type FTP [File Transfer Protocol] commands by hand and you had to be able to do address-mapping in your head between IP addresses and host names and you had to know where all the FTP archives were; you had to understand IRC [Internet relay chat] protocols, you had to know how to use this particular news reader and that particular Unix shell prompt, and you pretty much had to know Unix itself to get anything done. And the current users had little interest in making it easier. In fact, there was a definite element of not wanting to make it easier, of actually wanting to keep the riffraff out.[19]

So this bored, talented, larger-than-life, Internet-obsessed hacker decided that it would be interesting to plug the gap by giving the Net an easy-to-use, easy-to-get graphical interface. Andreessen wanted to let the riffraff in.

He approached a friend called Eric Bina to join him in this crusade. Bina was a formidable programmer, famous within NCSA for his ability to write robust code for several days without a break, and for never being able to resist technical challenges. But Bina was a full-time NCSA staffer and Andreessen was an undergraduate intern with zero official status, so nothing might have happened had not Joseph Hardin, Bina's (and indeed Andreessen's) manager, also sussed the significance of the Web around the same time. Hardin felt that a graphics-powered browser would be a good project for NCSA and so he gave the go-ahead for the project.

Mosaic was created in an astonishing burst of creativity, starting in January 1993 and lasting for just under three months. Bina and Andreessen would work three to four days straight, then crash for about a day. Working round the clock, arguing about all kinds of computing and non-computing issues at an all-night café, the two programmers achieved Gilbert and Sullivan-type synergy. 'We each did the job that most appealed to us,' Bina later told George Gilder, 'so each of us thinks the other did the hard stuff.' Bina wrote most of the new code – in particular, the graphics, modifying HTML to handle images, adding a GIF[20] decoder and colour management tools. Like all good programmers, he did it by adapting software tools which already existed – particularly a toolkit well known to UNIX programmers called Motifs. Andreessen's contribution was to take apart the library of communications code provided by CERN and rewrite it so it would run more quickly and efficiently on the network. Between them the two wrote Mosaic's 9,000 lines of code, in the process producing the most rapidly propagated piece of software ever written.

It was clearly an extraordinary partnership based on the fusing of complementary talents. In many respects the two seemed to be polar opposites of one another. Bina is small, cautious, focused, economical, reclusive; Andreessen is physically (very) large, daring, expansive, prodigal, go-getting. And yet they formed one of the most formidable programming partnerships since Thompson and Ritchie created UNIX.

As the project gathered steam, others at NCSA were drawn in. Because Andreessen and Bina were writing their code on UNIX machines, their browser was likely to wind up as yet another software toy for the Net's elite users unless it was translated to run on ordinary PCs and Macintoshes.[21] And porting it to mass machines was central to Andreessen's vision. The whole point of

Mosaic, in his view, was to empower non-geeks to use the Net. Jon Mittelhauser, a graduate student, took on the job of porting the software to the PC; one of his great contributions was the idea of having the mouse pointer change shape into a hand when it passed over a hotlink on a Web page. A Yugoslav called Aleks Totic (aided by another student, Mike McCool) volunteered to convert the program to run on the Apple Mac. McCool's twin brother Rob wrote the server software which the team eventually released along with Mosaic.[22]

On 23 January 1993, the following message appeared in several specialist Usenet conferences.

From: Marc Andreessen (marca@ncsa.uiuc.edu)

Sat, 23 Jan 93 07:21:17–0800

By the power vested in me by nobody in particular, alpha/beta version 0.5 of NCSA's Motif-based networked information systems and WorldWide Web browser, X Mosaic, is hereby released.

Cheers,

Marc

What Andreessen was doing was signalling to the Internet community that the software was now available for downloading across the network. Having posted the message, he then sat back to monitor the log automatically kept by the NCSA server as it responded to download requests. Within ten minutes of first posting the message, someone downloaded Mosaic. Within half an hour, a hundred people had it. In less than an hour Andreessen was getting excited e-mail from users all over the world. It was the Net equivalent of that moment when Mission Control says 'We have liftoff.'

Thereafter the UNIX version of Mosaic spread like wildfire through the worldwide computing community. The Mac and PC versions followed shortly afterwards. Within a few months, it was estimated (nobody at that stage was keeping precise records) that the number of downloads numbered hundreds of thousands. Objective measures of the impact of Mosaic also began to emerge. For example, in March 1993 – just a month after the official release of the Alpha version of the UNIX browser, Web traffic accounted for just 0.1 per cent of the traffic on the part of the Net known as the NSF backbone.[23] By September, there were over 200 known servers and the Web accounted for 1 per cent of backbone traffic – that is, it had multiplied tenfold in little over five months.

Mosaic was not the first browser, but it was the one which captured the market and shaped the future. This was partly due to the fact that it ran on simple desktop machines rather than fancy UNIX workstations. It also had something to do with the fact that it was the first browser which looked like a piece of modern, personal computer software: it had things like buttons and scroll bars and pull-down menus.

But perhaps the most significant thing about Mosaic was that it was designed to interpret a new HTML element – <IMG>, the image tag. In doing so it allowed Web pages to include images side by side with text for the first time, thereby making them potentially much more attractive to the legions of people who would be turned off by slabs of hypertext. At the time, the decision to extend HTML to handle images in this way was controversial in some quarters, mainly because image files tend to be much bigger than text files. A full-colour A4 picture, for example, runs to dozens of megabytes.

Many people saw the transmission of images as a wasteful consumption of scarce bandwidth and issued dire predictions of the Net being choked to death by the transmission of millions of packetised pictures. Even Bina and Andreessen argued about it. Andreessen maintained that people would not flood the Net with frivolous image files; and even if they did, he claimed that the system would increase its capacity to handle the load. Bina thought people would abuse it. 'I was right,' Bina says now. 'People abused it horribly . . . But Marc was also right. As a result of the glitz and glitter, thousands of people wasted time to put in pretty pictures and valuable information on the Web, and millions of people use it.'[24]

Tim Berners-Lee was also very critical of the addition of the <IMG> tag. Andreessen recalls being 'bawled out' by him in the summer of 1993 for adding images to the thing. The frivolity that the visual Web offered worried its inventor because 'this was

supposed to be a serious medium – this is serious information'.[25]
What this exchange portended was the change in perspective – the
paradigm-shift if you will – which was to fuel the Web's phenomenal growth from that point onwards. Berners-Lee and his colleagues saw their creation as a tool for furthering serious research
communications between scientific researchers. The kids at NCSA
were more pragmatic, less judgemental. 'Academics in computer
science', said Jon Mittelhauser to Reid,

> aren't always motivated to just do something that people want to
> use. And that's definitely the sense we always had of CERN. And I
> don't want to mischaracterise them, but whenever we dealt with
> them, they were much more interested in the Web from a research
> point of view, rather than a practical point of view . . . The concept
> of adding an image just for the sake of adding an image didn't make
> sense [to them], whereas to us, it made sense because face it, they
> made pages look *cool*.[26]

This is an over-simplification of the CERN perspective if only
because (as we have seen) one of the key elements in their project
from the outset was the creation of a line-mode browser which
would make the Web accessible to almost anyone with a Net
connection. But Mittelhauser still puts his finger on the important
difference between the two groups. Berners-Lee was like Thomas
Edison, who thought that the phonograph he'd invented was for
dictating office memos; Andreessen & Co. were closer in spirit to
Emile Berliner, the guy who realised that the Killer App for the
phonograph was playing pre-recorded popular music.[27] And, as it
happened, it was the NCSA lot who had the future in their bones.

After Mosaic appeared, the Web went ballistic. The program spread
like wildfire across the world. As it did so, the numbers of people
using the Net began to increase exponentially. As the number of
users increased, so also did the numbers of servers. And as people
discovered how simple it was to format documents in HTML, so the
volume of information available to Web users began to increase
exponentially. It was a classic positive feedback loop.

The fallout from this explosion is clearly visible in the statistical data collected by Matthew Gray at MIT which show the traffic over the NFS Internet backbone broken down by the various protocols (see table). What this table shows is that in two years the volume of Internet traffic involving Web pages went from almost nothing to nearly a quarter of the total.

Date	% ftp	% telnet	% netnews	% irc	% gopher	% email	% web
Mar.93	42.9	5.6	9.3	1.1	1.6	6.4	0.5
Dec.93	40.9	5.3	9.7	1.3	3.0	6.0	2.2
Jun.94	35.2	4.8	10.9	1.3	3.7	6.4	6.1
Dec.94	31.7	3.9	10.9	1.4	3.6	5.6	16.0
Mar.95	24.2	2.9	8.3	1.3	2.5	4.9	23.9

Source: www.mit.edu/people/mkgray/net/web-growth-summary.html.

The spread of the Web was like the process by which previous communications technologies had spread – but with one vital difference. It's a variant on the chicken and egg story. In the early days of the telephone, for example, people were reluctant to make the investment in the new technology because there were so few other people with telephones that it was hardly worth the effort. The same was true for electronic mail. 'Who would I send e-mail to?' was a common lament from non-academics in the early days of electronic mail. But, once a critical mass of other users in one's own intellectual, occupational or social group had gone online, suddenly e-mail became almost *de rigueur*.

The one great difference between the Web and, say, the telephone was that whereas the spread of the telephone depended on massive investment in physical infrastructure – trunk lines, connections to homes, exchanges, operators, engineers and so forth – the Web simply piggy-backed on an infrastructure (the Internet) which was already in place. By the time Mosaic appeared, desktop PCs were ubiquitous in business and increasingly common in homes. In the US in particular the existence of vast online services like America Online and CompuServe meant that modems had also begun to penetrate the home as well as the business market. The world, in other words, was waiting for Mosaic.

The one outfit which wasn't ready for Mosaic, paradoxically enough, was NCSA. As the program spread, the organisation found itself overwhelmed by what its students and hackers had released. NCSA, explained Andreessen afterwards, was a research institute, not a company set up to market and support software.

What was happening, of course, was that the Web tail had begun to wag the supercomputing dog. Despite its many virtues, Mosaic was not the easiest program to load and operate. It made computers crash. And if you clicked on a link to a server which was temporarily swamped, or offline and unable to respond to your request, your machine simply froze, leaving you with the sole option of rebooting and perhaps losing valuable data. For many users the idea of the Web was so novel that they tied themselves in knots trying to get started. They began besieging the NCSA for help, and the Center found itself swamped. And not just by people seeking help, either. Many of the calls were from people asking how they could get their hands on a copy of the program. Andreessen remembers getting calls asking, 'What do we need to run it?', and even a couple of calls from people wondering if they needed a computer to use Mosaic.

As the pressure increased, commercial pressure on its young developers also mounted. The NCSA's mission included the licensing of its inventions to commercial companies, but when it did so its staff saw none of the royalties. 'Companies started to come to us,' said Andreessen. 'They were saying: "Let us have it, how much do we pay? We'll give you money!"'

Mosaic's creators were thus experiencing many of the demands of working in a commercial company – providing 'customer' support, for example – but receiving none of the rewards which normally accompany such pressure. The discontent was fuelled by another development which would not have surprised anyone who knows anything about the dynamics of bureaucracy. While Mosaic was being written, most of the NCSA management neither knew nor cared about it. Why should they? Their core business, after all, was the management of a supercomputer site. But then Mosaic hit the streets and the management sat up and took notice. Suddenly what a small group of guerrilla hackers had wrought became *an NCSA achievement*, confirming the old adage that while failure is invariably an orphan, success has many fathers.

Given that the decline in supercomputing represented a real threat to NCSA's future, Mosaic must have seemed like manna from heaven to the Center's management. 'They were raking in millions of dollars per year in federal money for supercomputing,' said Andreessen, 'and no one really wanted to use supercomputers any more, so they sort of had two alternatives. One was to give up the federal funding, and one was to find something else to do. If you're an academic, you're certainly not going to give up your grant. So you look for something else to do. So they figured out that this was it fairly quickly.'[28]

Predictably, the management's Pauline conversion grated on the Mosaic boys. 'Suddenly,' recalled Jon Mittelhauser,

> we found ourselves in meetings with forty people planning our next features, as opposed to the five of us making plans at 2:00 A.M. over pizzas and cokes. Aleks [Totic], who had basically done the Mac version, suddenly found that there were three or four other people working on it with him, according to NCSA. And they were like his bosses, telling him what to do and stuff. And how can I put this politely? We didn't think that any of them had Aleks's ability or foresight.[29]

In the circumstances, the surprising thing was not that the Mosaic team left NCSA, but that they stayed as long as they did.

Andreessen left NCSA in December 1993 with the intention of abandoning Mosaic development altogether. He moved to Mountain View, California, a little town just south of Palo Alto, and took up a job with a small software company, Enterprise Integration Technologies, which was engaged in developing Web security products. This is where he was in February 1994 when, out of the blue, he received an e-mail message from someone called Jim Clark. 'You may not know me,' it began, 'but I'm the founder of Silicon Graphics . . .'[30] As it happened, Andreessen did know him for the simple reason that everyone knew who Jim Clark was. He was already a celebrity in the industry as the man who made the machines which came highest on every hacker's wish list – Silicon Graphics workstations.

Clark had come to business via the scenic route, going into the Navy after high school, then working his way through college and getting a doctorate in computer science in 1974. He had then taught at Stanford, specialising in computer graphics. On the floor of the lab above him were the guys who founded Sun Microsystems. In the basement of the same building were the people who founded Cisco, the company whose routers still pass packets across most of the Net. In 1981, Clark followed them on to the classic Silicon Valley trail: he left Stanford with a cohort of his best students and founded a company making dream machines. In less than a decade, Silicon Graphics workstations were being used to design special effects for Hollywood movies and to lure smart graduate students like Marc Andreessen into research labs.

In the time-honoured fashion of many charismatic entrepreneurs, however, Clark had grown disenchanted with his corporate child as it metamorphosed into a mature organisation. He and Silicon Graphics eventually parted company in January 1994 when Clark walked out the door with a large pile of cash and a feeling that he wanted to do something new. It was around that time that Bill Foss, once described as 'a midlevel marketing grunt' who had worked with Clark for years, showed him Mosaic and blew his mind. He had found what he was looking for. And shortly after that he found the man who had made it happen.

Thereafter, events moved quickly. Clark had the money, the entrepreneurial know-how and the contacts; Andreessen knew the product, and the people who could make an even better one. The company which was eventually to become Netscape[31] was formed, and the pair flew to Champaign-Urbana, checked in at the University Inn and invited all of the original Mosaic team to come and see them. They were all offered good salaries and 1 per cent of the stock in the new company. By the end of the day, seven (including Eric Bina) had signed up. Lacking a printer, Clark typed a letter of appointment on his laptop and then faxed it seven times to the fax machine in the hotel lobby. The others then signed on the dotted line. The deal was done.

Their mission was to produce an improved Mosaic, but rumblings of an intellectual-property suit from NCSA convinced them they needed to write a new browser from the ground up without using a line of the original code. This was a blessing in disguise, of course, because Mosaic had been written to run over high-speed, permanent Net connections, whereas the bulk of new Web users had only dial-up modem connections. And they were the market of the future.

Andreessen worked his new team harder than they had ever worked in their lives. He and Clark were driven by a vision and haunted by a nightmare. The vision was that the first company to tap into the unfathomable depths of the Web would make a fortune; the nightmare was that NCSA licensees would get there before them. (On the day the seven resigned from the Center, it had licensed Mosaic to a company called Spyglass.)

Clark's and Andreessen's new company was formally incorporated on 4 April 1994. The new browser, called Netscape Navigator 1.0, was launched in late December after the kind of programming marathon known in the industry as a 'death-march' – writing and debugging code eighteen hours a day, seven days a week; surviving on junk food and fizzy drinks; sleeping on the floor; putting personal hygiene on the back burner. Even then it was a damned close-run thing: the company almost ran out of cash in December and Clark saved it only by firing fifteen people.

But the effort was worth it: the new browser was indeed a better Mosaic. It was cleaner, more secure, and had support for more elegant layouts and more elaborate documents. It was also faster: I can still remember the shock of discovering that Navigator ran faster on my home computer's 14,400 bits-per-second modem than Mosaic did across the campus Ethernet at work.

Reckoning that the real money was to be made from selling the server software to go with the program, Clark and Andreessen decided to give Navigator away[32] on the Net. It was their variation on the old business theme: give the razors away, but remember to charge for the blades. Six million copies of the program were downloaded under these liberal terms. And lots of companies paid for multiple licences to put the browser on all their desktops: this trade, together with the revenue from sales of the server software, grossed $365,000 in the first two weeks after Navigator was released. In the second quarter of trading the company took nearly $12 million in telephone sales – many of them to companies seeking to set up *intranets* – their own, in-house mini-internets. In four months Netscape's share of the global browser market went from zero to 75 per cent, which must be the fastest growth in market share in the history of capitalism.[33]

In choosing to use the Net as their distribution vehicle, Andreessen and Clark had also flipped the computing industry into fast-forward mode. Previously, the rate of diffusion of new products had been limited by the inertia and lags of the manufacturing and distribution chain. But, by dispensing with all the packaging, warehousing, shipping and advertising that goes with a conventional product, Netscape had effectively launched an era when you could finish a product one day and have hundreds of thousands of users the next. The old era of two-year product cycles was over.

The implications of this were incalculable. For one thing, it meant that the pace of innovation had irreversibly been ratcheted up. From now on, time in Cyberspace would be measured in 'Web years', each one of which was equivalent to about seven offline ones. At this frenetic pace it was only a matter of time before people started to suffer from change-fatigue.

Much the same applied to competition. Without the need to assemble a huge manufacturing and distribution operation, just about anyone could get into your market – and gain a sizeable share of it in the blink of an eye. All of which meant that from now on there were only going to be two kinds of successful operations on the Net – the quick and the dead.

The import of this was not lost on Wall Street. On 9 August 1995, Netscape Communications Inc. was floated on the stock market, with its shares priced conservatively at $28 apiece. They started at $71, went as high as $74.75 before closing the first day's trading at $58.25. Jim Clark's holding in the company was suddenly worth $566 million. Andreessen, for his part, found himself worth $58 million at the age of twenty-one. He seems to have taken it in his stride. 'I was at home in bed,' he told *Time* magazine. 'I had been up until, like, 3 in the morning, working, so I woke up at 11, logged in from home, looked at Quote.Com. My eyes went . . . [He makes a face of astonishment] . . . Then I went back to sleep.'[34] The following day a slightly stunned *Wall Street Journal* commented that, while it had taken General Dynamics forty-three years to become a corporation worth $2.7 billion, Netscape Communications had achieved the same thing in about a minute.

16:
Home sweet home

The road of excess leads to the palace of wisdom.
William Blake, 'The Marriage of Heaven and Hell', (1790–3)

It was too good to last, of course.

For a time, Netscape had the field to itself. The company grew like crazy. By December 1995 its stock had reached $170 a share. By the next summer pedestrians couldn't walk on the pavement near the company's loading dock because of the press of delivery trucks. Its workforce had tripled to nearly 2,000 people. Its product had become the *de facto* world standard. Andreessen had metamorphosed into a kind of high-tech pin-up boy. He was on the cover of every news and business magazine. People talked about him as the next Bill Gates. Wall Street hung on his every word. *Fortune* magazine called him 'the hayseed with the know-how'.

This was heady enough stuff for an individual, but for his company it was like snorting premium-grade cocaine. A perceptible whiff of hubris began to permeate its public relations. Netscape executives began to tell anyone willing to listen that the browser – that is to say Netscape's browser – would be the key to the future of computing. There would come a time, they said, when the first thing people would see when they booted up was a browser. They pointed to products like the CD-ROM version of *Encyclopaedia Britannica*, which launched straight into Netscape and used a query form just like AltaVista or any other Internet search engine. And they cited the second release of their product – Navigator 2.0 – and the way it enabled the user not only to browse the Web, but also to send and receive e-mail and participate in Internet news groups. It provided everything the online consumer needed in one software package. This was the way it was going to be, said Andreessen and his guys – the browser would eventually become the operating system.

To some extent this fine talk was just bravado, designed to conceal the gnawing fear in their vitals. From the outset, Clark and Andreessen had known that some day Microsoft, the 800-pound gorilla of the computing jungle, would come looking for them. What astonished them at the beginning was that the monster seemed to be asleep.

One of the great mysteries of modern corporate history is why Microsoft took so long to suss the Net. Here was a company employing hundreds of the smartest people in the computer business and yet it appeared blind, deaf and mute in the face of the most revolutionary change in its environment since the invention of the microprocessor. There was a time when my colleagues and I gaped at this wilful blindness. How could they miss it, how *could* they? The Net was staring Bill Gates in the face – and yet he didn't seem to *get* it. It looked as though history was repeating itself. It wasn't all that long ago since IBM – that mighty, overweening corporate brute we called 'Big Blue' – had overlooked the PC until it was too late. Gates had often commented in public about this fateful omission. Now Microsoft had become the IBM *de nos jours* – and it looked as though it too was sleepwalking to disaster.

One charitable explanation has it that Gates and his team were so distracted by the anti-trust actions launched by the US Department of Justice that they took their eye off the Internet ball. Another – less charitable – interpretation is that they thought of the Net as just a kind of gigantic version of CompuServe – that is to say, something they could buy out or wipe out when the time came. A third theory attributes the blind spot to Microsoft's relative isolation in Seattle.

James Wallace, one of Gates's biographers, claims that Microsoft did not have an Internet server until 'early 1993', and that the only reason the company set one up was because Steve Ballmer, Gates's second-in-command, had discovered on a sales trip that most of his big corporate companies complained about the absence of TCP/IP facilities in Microsoft products. Ballmer had never heard of TCP/IP. 'I don't know what it is,' he shouted at a subordinate on his return. 'I don't want to know what it is. [But] my customers are screaming about it. Make the pain go away.'[1]

Whatever the explanation for Microsoft's blind spot, the record shows that some time in February 1994 Steven Sinofsky, one of Gates's closest technical advisers, went to Cornell University and encountered legions of students drooling over Mosaic. When he got back to HQ, a thoughtful Sinofsky booked time with his boss and gave him the Mosaic treatment.[2]

Then Gates got it. Instantly. He convened a conference of senior technical managers for 5 April (the day after Netscape was incorporated!) and afterwards issued a stream of memos on what he called 'The Internet Tidal Wave' which had the effect of turning his entire company on a dime. He talked about a 'sea change', of the need to develop a browser and of the imperative to incorporate the Internet into *all* the company's products. That included not just its bestselling applications like Word, Excel, PowerPoint and Access, but also the next version of its operating system, Windows 95.

The trouble was that at the time Microsoft didn't have a browser. So it did what it has traditionally done in such circumstnaces – it went straight out and bought one. In this case, it went to Spyglass, the company which had licensed the Mosaic software from NCSA. The rebadged product was christened Internet Explorer 1.0 and distributed free over the Net, on computer magazine cover disks and via every other outlet Microsoft could think of. And of course it was bundled with every copy of Windows 95 from the moment the new operating system was launched in August 1995.

For Netscape, the launch of Internet Explorer was the ultimate reality check. The glorious dream which had ignited on 9 August 1995 was over. The gorilla had arrived on the block and he wasn't going to stand any nonsense. On Pearl Harbour day, 7 December 1995, Gates made a speech to computer industry analysts and commentators. The sleeping giant had awoken, he told them, consciously evoking the moment in 1941 when the Japanese Air Force had made the terminal mistake of rousing the United States from *its* slumbers. He outlined his new Internet strategy and explained that the company was developing versions of its browser for the Apple Macintosh and Windows 3.1 – and that it would be giving them away.

What newspapers took to calling the 'Browser Wars' had begun. It was the computing equivalent of those illegal heavyweight bouts in which the fighters go on until one of them dies from punishment or exhaustion. Netscape would launch a new version of its browser. Microsoft would counter with a new release of its program. And so it went on through versions 2, 3 and 4 of both browsers.

In this kind of slugfest though, stamina is all; and stamina equates to resources, and everyone knew whose pockets were deepest. Microsoft had a mountain of cash, no debts and a fantastic revenue stream. Gates ran his company on the basis that it could survive for an entire year without earning a cent. All other things being equal, therefore, Microsoft was bound to win.

It looked like the end-game had begun. Netscape stock started to slide. The company began to talk about alternative strategies, about leveraging its hold on the server market, about becoming a 'group-ware' company or even a 'portal' – that is, a major destination or jumping-off point – on the Web. Its executives began to look for a way out of a browser war they felt they could not win.

They embarked on a campaign to reinvent their company – reorienting it to concentrate on the corporate intranet market, Web commerce and other potentially lucrative businesses. By the time Version 4 of their software arrived, it was as bloated as anything Microsoft had ever produced.[3] Now called Communicator, it of course included a browser – but only as one element in a whole suite of programs for real-time conferencing, e-mail, diary manage-ment, scheduling meetings, HTML authoring and other corporate tasks. 'Forget the browser,' was the implicit message, 'we've moved on to higher things.'

As a corporate strategy, it had the merit of avoiding confronta-tions Netscape was bound to lose. But implicit in it was an admission that the day was coming when the company's software would no longer be the *de facto* standard for browsing the Web. And that outcome would be determined, not by the quality of the competing software, but by Microsoft's monopolistic grip on the world's desktops.

It was unfair and unjust. After all, it was Andreessen, not Gates, who had first seen the potential of the Web, and turned it into the mother of all business opportunities. But, as Gates might well have said, what's fairness got to do with it? This is business, stoopid.

Andy Grove of Intel once observed that, in the computer business, 'only the paranoid survive'. This is the Microsoft mantra. No matter how big the company became, no matter how dominant it was in its chosen markets, Gates and his lieutenants were forever looking over their shoulders, waiting for someone to do to them what they had once done to IBM.

Microsoft launched Internet Explorer 3.0 in the summer of 1996. It got good reviews – indeed some commentators thought it superior to the corresponding version of Navigator. Microsoft continued to take browser market share from Netscape. But not, it seems, at a rate acceptable to a paranoid management. On 20 December 1996, a senior Microsoft executive, Jim Allchin, dispatched a routine e-mail to his boss, Paul Maritz. Its subject was 'concerns for our future'. 'I don't understand', wrote Allchin, 'how I[nternet] E[xplorer] is going to win. The current path is simply to copy everything Netscape does packaging and product wise . . . My conclusion is that we must leverage Windows more. Treating IE as just an add-on to Windows which is cross-platform is losing our biggest advantage – Windows marketshare.'[4]

As a business argument, this made excellent sense. The only trouble was that it is against the law in the US for a company to seek to leverage (that is, exploit) a monopoly product (like the Windows operating system) in order to gain a competitive edge in another market. Later, Allchin protested that he hadn't meant anything illegal by his e-mail. Perish the thought. What he was trying to say, he wailed, was Microsoft should be 'leveraging Windows as a matter of software engineering to build a better project'.

Ho, ho. Unfortunately for Allchin, the US Department of Justice

did not see it that way. The reason we know about his e-mail
message is that the DoJ launched an anti-trust suit against Microsoft
in late 1997 and it was one of the exhibits dredged up by a trawl
through the company's files. And while the mindset it revealed may
have caused some raised eyebrows in Washington, to the folk at
Netscape it merely confirmed what they had known all along. Bill
Gates only knows how to play one game. Americans call it hardball.

The anti-trust suit gave Netscape a welcome breathing space. It
looked as though the DoJ action would distract Gates's attention for
a while. But nobody in his right mind would regard it as anything
more than a temporary respite. Anti-trust suits had a habit of
dragging on for years. The case was complex, involving not just
competitive and anti-competitive practices but also the intricacies
of operating system and browser design. Microsoft was contesting it
vigorously, and had the resources and determination to fight all the
way to the Supreme Court. Indeed it was said that Gates had once
threatened to move his entire operation to some offshore jurisdic-
tion with less stringent anti-trust concerns where he could do as he
damn well pleased. And with his kind of money he could buy a
small country.

In the meantime Internet Explorer 4.0 was on the streets. By
ticking a box during the installation process, you could cause it to
integrate seamlessly with your Windows desktop. For Netscape the
central problem remained: how could they give their product a
chance of remaining the *de facto* standard when they were no
longer the dominant player in the market? The answer, of course,
was staring them in the face, though it took them ages to tumble to
it. And the funny thing is that, in the end, they got it off the Net, in
an essay written by a hacker of whom most Netscape board
members had probably never heard. His name was Eric Raymond.

Eric S. Raymond is an ebullient, rumpled, slightly messianic
figure who looks and sounds like the media stereotype of a
computer hacker. Within the computer business he was known
originally for *The New Hacker's Dictionary*, an indispensable bible
which he conceived and published on his own. He is a gifted
programmer who describes himself as 'a neo-pagan, libertarian,
arrogant son of a bitch' with a 'Napoleon complex' and a penchant

for semi-automatic weapons. Close inspection of his writings suggests a deep suspicion of Microsoft and all its works and pomps. He suffers from cerebral palsy, a condition that encouraged him to look upon computers as something over which he could exercise the kind of control denied him in the physical world. Raymond's essay had the enigmatic title 'The Cathedral and the Bazaar', and it was published only on the Net.[5] Its subject was the question of how to create complex software that is stable and dependable.

This is one of the great questions of our time. We live in a world increasingly dependent on software (as the Millennium Bug crisis cruelly revealed), and yet we have great difficulty producing computer programs that are bug-free and reliable. This was not so much of a problem in the early days when software had to be compact in order to fit into the limited memory space of early computers. But nowadays RAM is dirt cheap and we are enslaved to programs like operating systems, browsers, air-traffic-control systems, word-processors even, which are fantastically complex, intricate, obese products – with millions of lines of code and billions of possible interactions.

And reliability is not the only problem which has grown with increasing complexity. Another is the difficulty of producing software to specification, on time and within budget. Creating huge programs like Windows NT or Communicator 4 effectively requires the industrialisation of programming, with large project teams, division of labour and heavy-duty project management. The trouble is that programming is an art or a craft, not a science, and it doesn't lend itself readily to industrialisation. Most programmers see themselves as craftsmen (or craftswomen) and behave accordingly. They do not take kindly to being regimented.

For decades, the canonical text on software project management was Frederick Brooks's *The Mythical Man-Month*.[6] Brooks headed the team which wrote the operating system for the IBM 360 range of mainframe computers and his book is a wonderfully elegant, thought-provoking reflection on what he learned from that harrowing experience. Its strange title, for example, comes from the observation that adding extra programmers (man-months) to a team that is falling behind schedule invariably results in the enlarged team falling even *further* behind. (The reason, which is obvious when you think about it, is that some of the time that should be spent programming goes into briefing the new guys and into the communications overhead of a bigger team.)

Brooks's approach to software production is what has ruled the industry for thirty years. His governing metaphor was the building of Reims cathedral. 'The joy that stirs the beholder', he wrote of that great edifice,

> comes as much from the integrity of the design as from any particular excellences. As the guidebook tells, this ingenuity was achieved by the self-abnegation of eight generations of builders, each of whom sacrificed some of his ideas so that the whole might be of pure design. The result proclaims not only the glory of God, but also His power to salvage fallen men from their pride.[7]

This metaphor – software produced by the self-abnegation of individual programmers in deference to an overarching, grand design – was what Raymond was attacking in his essay. It was the 'cathedral' of his title. His argument was that the approach was doomed to failure as software became more complicated, and that its limitations were being more cruelly exposed with every succeeding release of operating systems and applications. Companies like Microsoft and Netscape could no longer control the monsters they were creating. Even more importantly, they could not ensure their reliability. However elegant the cathedral metaphor might seem in theory, when it came to the gritty practice of producing complicated pieces of software, the approach simply didn't work. There had to be a better way.

There is. It's the Open Source approach. 'The central problem in software engineering', Raymond later told an interviewer,

> has always been reliability. Our reliability, in general, sucks. In other branches of engineering, what do you do to get high reliability? The answer is massive, independent peer review. You wouldn't trust a scientific journal paper that hadn't been peer reviewed, you wouldn't trust a major civil engineering design that hadn't been indepen-

dently peer reviewed, and you can't trust software that hasn't been peer reviewed, either. But that can't happen unless the source code is open. The four most critical pieces of infrastructure that make the Internet work – Bind, Perl, Sendmail and Apache – every one of these is open source, every one of these is super reliable. The Internet would not function if they weren't super reliable, and they're super reliable precisely because throughout their entire history people have been constantly banging on the code, looking at the source, seeing what breaks and fixing it.[8]

The metaphor Raymond chose to capture the essence of the constructive free-for-all that is the Open Source movement was the bazaar. The greatest example of its power is probably Linux, the UNIX-inspired free operating system which spread like wildfire through the computing world because of its remarkable reliability and stability.[9] In his essay, Raymond drew extensively on his own experience of developing a piece of e-mail software within the Open Source tradition, speculated on why people working in this tradition are so co-operative and outlined some maxims about how to create great software and avoid the traps of the cathedral approach.

'The Cathedral and the Bazaar' is one of those seminal documents that appear once in a generation and articulate what thousands of people have been thinking but never managed adequately to express. I remember reading it for the first time and thinking that Raymond had managed to distil the wisdom of the Net. This is the greatest co-operative enterprise in the history of mankind, he was saying. There is no intellectual problem on earth it cannot crack. Its collective IQ is incalculable. So why don't we use its power?

Someone at Netscape – we don't know who – saw Raymond's essay, and passed it round. Some board members – again we don't know how many – read it. And as they brainstormed their way to a corporate survival strategy, the force of Raymond's argument bore down on them until in the end they saw the point. They decided to set their browser free – to release the source code under an Open Source licence and let the Net do the rest.

On 22 January 1998 Netscape Communications Corporation

issued a press statement announcing its decision to make the source code for the next generation of Communicator available for free licensing on the Internet. 'This aggressive move', said the company, 'will enable Netscape to harness the creative power of thousands of programmers on the Internet by incorporating their best enhancements into future versions of Netscape's software.'

'By giving away the source code for future versions', said Jim Barksdale, Netscape's President and Chief Executive Officer,

> we can ignite the creative energies of the entire Net community and fuel unprecedented levels of innovation in the browser market. Our customers can benefit from world-class technology advancements; the development community gains access to a whole new market opportunity; and Netscape's core businesses benefit from the proliferation of the market-leading client software.

Afterwards, they threw a party at some joint in Mountain View, the town where Andreessen had settled after he left NCSA and where Jim Clark found him. As people danced and drank and talked, a giant back-projection screen displayed the strangest sight – incessantly scrolling lines of typescript, millions of lines of it, mostly incomprehensible. It was, of course, the code. Or, as hackers say, the Source.

I spent most of that day in London at a seminar and knew nothing about the decision. Afterwards I had dinner with a friend. We talked about the Net, among other things, and about this book. I got the last train back to Cambridge and arrived exhausted, acutely conscious of the fact that I had a newspaper column to finish before I could get to bed.

As I sat down to write, my e-mail system signalled that it had something for me. It was a two-line message from my dinner companion. The first was the URL of the press statement. On the second line there were three words: 'Netscape's coming home!' I sat there in the dark, digesting the decision that Barksdale, Andreessen & Co. had made, stunned by the enormity of what they had done. And what came to mind then had nothing to do with share prices or Internet futures or corporate strategy or all the other things that drive Silicon Valley and the industry, but a fragment of T. S. Eliot's poem, 'Little Gidding'. 'We shall not cease from exploration,' it reads,

> And the end of all our exploring
> Will be to arrive where we started
> And know the place for the first time.

EPILOGUE:

The wisdom of the Net

Man is a history-making creature who can neither repeat his past
nor leave it behind.

W. H. Auden, 'The Dyer's Hand', 1963

This is not, as Churchill remarked in another context, the
beginning of the end; but it is the end of the beginning.
When people learned I was writing a book about the history of the
Net they were often incredulous, not just because they thought the
network wasn't old enough to have a history, but because they
regarded it as somehow absurd that one should try and pin down
something so fluid. Was it not the case that almost anything one
wrote would be out of date by the time it was published? And that
any attempt to call a halt to the story would seem arbitrary in
retrospect?

Yes to both. No sooner had I completed the previous chapter
about Netscape and the Open Source movement, for example, than
Netscape was bought by AOL – with lord knows what long-term
consequences. And, even as I write, serious commentators are
beginning to talk about the break-up of Microsoft as it becomes
abundantly clear that the 800-pound gorilla of the computing
business has screwed up its defence against the trustbusters of the
US Department of Justice. Writing about the Net is like skating on
quicksand.

Similarly, writing about its future is a mug's game. When people ask me what will happen next I tell them to buy a crystal ball. If they persist, I invite them to think about it in navigational terms. Imagine, I say, you are sailing a small yacht. The outboard motor has broken down, so you are at the mercy of the wind and tides. The wind is blowing at ten knots from the south and the current is running due east at ten knots. Which way does your boat drift? And how fast?

Engineers (and presumably sailors) have a simple way of answering these questions. The wind and the current are both forces acting in particular directions. We call them *vectors* because they have a magnitude and a direction. Take a sheet of squared paper and mark a point on it. Draw a straight line of length ten units going due north and mark the end with an arrowhead to indicate the direction of the wind. This is the wind vector. From the top of the arrowhead now draw another line, also of length ten units, but this time at right angles heading due east. Put another arrowhead on this to indicate direction. This is the current vector. Now draw a line from the starting point to the end of the current vector and you've got the direction your boat will drift. The length of the line indicates the speed at which it will move. It's what engineers call the *resultant* of the two forces acting upon it.[1]

This resolution-of-forces model is all you need to predict the future of the Net. The only difference is that there are now three forces driving the yacht. One is the human thirst for information. Another is the demand for entertainment and diversion. And the third is the motivation to make money from providing online goods and services. So the model has to operate in three dimensions rather than in two, but the principle is exactly the same. The Net will evolve along the resultant of these three vectors. Easy, isn't it? Alas, no. The problem is that we have no idea of either the magnitude or the direction of the vectors. We don't know the way the winds blow, or how fiercely. So we cannot draw the diagram, resolve the forces and find out where we're headed.

Let us therefore focus instead on something much more tangible – what we might learn from the past. What does the story of the Net tell us that might be useful in thinking about the present? First of all, it tells us something about the importance of dreams. Engineers and computer scientists and programmers are supposed to be hard-nosed, firmly earthed folk. And yet the history of the Net is littered with visionaries and dreamers who perceived the potential of the technology and tried to envision what it might enable us to do.

One thinks, for example, of Norbert Wiener, who was one of the first to realise that computers were more, far more, than calculators, and began to fret about the relationship humans would have with these machines. Or of J. C. R. Licklider, who saw them as workmates and thought of the relationship as symbiotic. Or of Vannevar Bush, who hoped they might save us from being engulfed by our own knowledge. Or of Douglas Engelbart, who decided to devote his life to finding a way of using computers to augment human capabilities. Or of Ted Nelson, who railed against forgetting and tried to create a universe of hyperlinked documents which would save us from collective amnesia. Or of Bill Atkinson, who wanted to put the power of the Macintosh into the hands of ordinary human beings. 'Where there is no vision, the people perish,' says the Old Testament.[2] The same holds for computer science. It was the visions of these men which drove the evolution of the ideas which led to the modern Internet. Without them its development would un-doubtedly have been different – and slower.

But it's not just enough to have visionaries. Someone Up There has to appreciate their importance and be prepared to back their hunches with someone else's money. This is where one begins to understand the importance of people like Bob Taylor – and Mike Sendall, Tim Berners-Lee's boss at CERN, for without them the visions of Engelbart & Co. would have withered for lack of resources. Taylor's special genius as an administrator lay in recog-nising talent and being prepared to back it. Early on in his career, during his time at NASA for example, he spotted Engelbart's importance when everyone else thought the Stanford researcher was off the wall, and started funnelling research funding his way. At ARPA, Taylor persuaded the agency's Director to give him a million bucks to create a network that nobody (least of all Taylor) knew how to build. And with the same unerring nose for talent, he identified Larry Roberts as the man to do it, and ruthlessly pressured the Lincoln Lab at MIT into sending him to Washington.

And the astonishing thing is that, after he left ARPA, Taylor did it again. In the late 1960s, the huge Xerox Corporation decided it had better find out about the 'paperless office' that Engelbart kept talking about. The company set aside a large pot of money, began building a research centre in Palo Alto and recruited Taylor to head up its Computer Science Laboratory. From his time at ARPA, he knew all the smartest engineers and computer scientists in the business, and he set about hiring them until it was said that of the top one hundred computer people in the US, Xerox PARC employed seventy-six.[3] Having recruited them, Taylor then let these guys get on with whatever they wanted to do. In the course of a few years, they repaid his trust by inventing the future. Most of the computing technology we use today – from the graphical user interface embodied in Microsoft Windows or the Apple Macintosh to laser printers and the Ethernet local area networking system – comes originally from that single laboratory.

What Bob Taylor understood better than anyone else since Licklider was that the way to get magic was to give clever researchers the freedom to think, while protecting them from demands for deliverables or deadlines or the constraints of the ordinary commercial world. His great fortune was to work for two outfits – ARPA and Xerox PARC – where he was able to make this happen.

He was also lucky to have lived in a more expansive age. Nowadays there are very few research labs – and no publicly funded agencies – which enjoy the freedom of action that Taylor could take for granted. Remember his story about getting a million dollars from his boss in twenty minutes? That would be inconceivable today: it would take half a dozen project proposals, an elaborate peer-review process, detailed scrutiny by the Office of Management and Budget and two years of waiting to get the contemporary equivalent of Taylor's million bucks. 'If the Internet hadn't existed,' one of his wizards told me in 1998, 'we couldn't invent it now.' And why? Because we live under the tyranny of the bottom line. The space which we were once happy to reserve for dreamers has been partitioned into offices for accountants.

*

Secondly, the story of the Net reminds one of the arbitrariness of history. The story of how this remarkable network came to be built is an exceedingly complicated one. But to relate it one must hack a narrative path through the jungle, and in the process do some injury to the truth. And it's not just we historians who are the vandals: as the generation that built the Net grows old, one can already see a certain amount of jostling for a place in its emerging pantheon. Some of them have taken to describing themselves as a 'father of the Internet' on their own Websites. And my own investigations into the background – which sometimes involved testing one interpretation against someone else's memory – prompted some vigorous e-mail exchanges between some of the characters in the story.

It's also interesting to see the way some people get written out, as it were. In 1998 Robert Cringely, who had previously published an excellent account of the PC industry,[4] managed to make a three-part television history of the Net without mentioning Paul Baran or Donald Davies – and indeed left most viewers with the distinct impression that packet-switching had been invented by Leonard Kleinrock rather than by them.

Similarly it's often forgotten that the version of the proposal which convinced CERN management to allocate resources to the World Wide Web project had Robert Cailliau as its co-author. Yet where is Cailliau now in the annals of the Web? And the conventional story of how the Internet evolved from the ARPANET gives the credit to Vint Cerf and Robert Kahn and mentions only in passing – if at all – the contributions made by John Shoch and his colleagues at Xerox PARC and the others who hammered out the TCP/IP family of protocols.

To some extent this airbrushing of the historical record is inevitable. For what is going on is the invention of a myth. And myths need heroes – single-handed slayers of dragons – rather than Network Working Groups, co-authors and other actors who might dilute the glory.

That said, it's impossible to read the history of the Net without being struck by the extent to which the genius of particular individuals played a crucial role in the development of the concept and its realisation in hardware and software. It's easy to look back now and laugh at Licklider's vision of 'man–computer symbiosis', or to describe Donald Davies's idea of packet-switching, Wesley Clark's notion of a subnetwork of message-processors, or Vint Cerf's concept of a gateway between incompatible networks as 'simple' or 'obvious'. But that, in a way, is a measure of their originality. They are those 'effective surprises' of which Jerome Bruner writes – the insights which have 'the quality of obviousness about them when they occur, producing a shock of recognition following which there is no longer astonishment'.[5] The fact that we live now in a packet-switched and networked world should not blind us to the ingenuity of the original ideas. And the people who conceived them should not be subjected to what the historian E. P. Thompson called 'the condescension of posterity'.[6]

This is not a fashionable view in some quarters, especially those lodged on the higher slopes of the history and sociology of technology. From that lofty vantage point, scientists, inventors and engineers look rather as rats do to a behaviourist – creatures running in a maze created by economic and social forces which they do not understand, and achieving success only when they press the levers which the prevailing order has ordained will yield success.[7] It's rather like the view famously expressed by Bertrand Russell that economics is about how people make choices and sociology about how they don't have any choices to make.

Under this view, the ARPANET would have emerged anyhow, even if Bob Taylor hadn't willed (and funded) it. The military–industrial complex needed a communications system capable of surviving a nuclear onslaught and would have obtained one eventually. Likewise, the multimedia conglomerates of this world needed something like the Web and would have created one even if Tim Berners-Lee had never been born.

It would be foolish to deny that this argument has some force. Computer networking would certainly have evolved even if the people who created the Internet had never existed, simply because the technological and economic imperatives to share computing resources were so strong. Similarly, a global multi-media distribution system would eventually have emerged from the marketing drives of the Disneys and Time-Warners and Murdochs who dominate the infotainment business.

But – and here's the critical bit – the networking systems that would have arisen from such institutional needs would have been radically different from what Licklider, Taylor, Berners-Lee and the others dreamed up and built. Does anyone seriously believe that a military–industrial complex left to its own devices would have consciously willed a network entirely devoid of central control, powered by a technology based on open standards which allows anyone to hook up to it? Or that Disney & Co. – the corporate behemoths who have built global businesses on *pushing* content *at* consumers – would have designed a 'pull' medium like the Web which enables anyone to become a global publisher and gives the consumer the power to fetch what he or she wants and nothing else?

This is the point that is continually missed by those who sneer at the Net because of its military provenance. They fail to appreciate the values that those who created it built into its architecture. 'It is no accident', writes Lawrence Lessig, the Harvard lawyer who understands the Net like no other member of his profession,

> that it came from the research communities of major universities. And no accident that it was pushed onto these communities by the demands of government. Once forced, researchers began to build for the Internet a set of protocols that would govern it. These protocols were public – they were placed in a commons, and none claimed ownership over their source. Anyone was free to participate in the bodies that promoted these commons codes. And many people did. There was a barn-raising . . . that built this Net.[8]

When Bill Clinton first ran for President, his chief political strategist, James Carville, advised him to focus obsessively on the state of the US economy as the core campaign issue and to ignore more or less everything else. Carville expressed this philosophy in a mantra which the entire campaign team was expected to recite at regular intervals: 'It's the economy, stoopid.'

Well, with the Net, it's the values, stoopid.

What are they? The first is it is better to be open than closed.

Lessig makes the point that there have been electronic networks since the late nineteenth century, but they were predominantly proprietary, built on the idea that protocols should be private property. As a result, they 'clunked along at a tiny growth rate'. He contrasts this with the Web, where the source code of every single page is open – for anyone to copy, steal or modify simply by using the View: Source button on their browser – and which is the fastest-growing network in human history: 'Nonproprietary, public domain, dedicated to the commons, indeed some might think, attacking the very idea of property – yet generating the greatest growth our economy has seen.'[9]

The openness which is at the heart of the Net is often portrayed as its central weakness by a corporate culture which cannot figure out how to make money from it. In fact, openness is the Net's greatest strength and the source of its power. It's why a system of such staggering complexity works so well, for one thing. It's also why the Open Source movement is significant – because it understands that the best way to produce high-quality, reliable computer software is to maximise the number of informed minds who scrutinise and improve it. As Eric Raymond once said, 'given enough eyeballs, all bugs are shallow'.[10] Or, as a software engineer said to Lawrence Lessig: 'The "ah-ha" for Open Source Software came to me when I realised, "Wait, open source is the Internet."'

The openness of the Net also applies to its future. The protocols which govern it leave the course of its evolution open. (Contrast that with France's legendary Minitel system – a flagship for its time, but now beached like a rusting Second World War destroyer while its erstwhile users struggle to get up to speed on the Net.) TCP/IP allows anyone to hook up to the Internet and do their own thing. It doesn't prescribe what they should do, only the lingo their software should speak. And it evolves to handle new applications and technologies.

Similarly HTML does not prescribe what people should put on their Web pages, and it evolves as an open standard to cope with their desires and demands for new facilities. Nobody plays God on the Net. Its protocols evolve in response to its users' needs and the possibilities offered by new technology – imperatives which are mediated through open institutions. As Dave Clark of MIT, one of the Net's Elders, observed, 'We reject: kings, presidents and voting. We believe in: rough consensus and running code.'[11]

The other core value of the Net is what Lessig calls 'Universal Standing'. Anyone can download the source code for the Linux operating system and alter it. You don't need to ask permission of the government, of Bill Gates or even of Linus Torvalds, who created the damn thing. And having downloaded and tampered with the code, you can put it back on the Net for others to do the same. Cyberspace is the only place I know where equal opportunity really is the order of the day. There is no formal power on the Net, no 'kings, presidents and voting', no agency which can say 'this is how it's gonna be because we say so'. But just because there's no formal power on the Net doesn't mean there's no *authority*. Lessig's principle of Universal Standing 'keeps open the space for individuals to gain – not power, but authority . . . One gains authority not by a structure that says, "You are the sovereign," but by a community that comes to recognise who can write code that works.'[12]

Everyone has a right to be heard, in other words, but not everyone is taken seriously. The Net, in this sense, is the ultimate meritocracy: 'It is the craft of your work, and the elegance and power of your solution, that commends it, and gives it power. Not your status, not your rank, not your corporate position, not your friendships, but your code. Running code, that by its power produces rough consensus.'[13] Most debates about the Net ignore all this, of course. In them, Cyberspace is regarded much as the African interior was regarded by European imperialists in the nineteenth century – as a region inhabited by 'lesser breeds without the law' with poor personal hygiene, little discipline and woeful ignorance of the laws of tort and contract, private property, the need for firm government and other appurtenances of civilisation. The talk is all about how the Internet should be regulated and controlled – of how, in other words, the procedures and norms of what we laughingly call the real world ought to be applied to the virtual one.

If there is a lesson to be drawn from my little history of this extraordinary phenomenon it is that we've got it the wrong way round. The real question is not what has the Internet to learn from us, but what might we learn from it.

Thomas Jefferson would have loved the Net. So would Thomas Paine. And so would my father, if he had lived to see it. The Web would have enabled him to be the world's greatest radio ham, without having to worry about passing the Wireless Telegraphy examination or putting up an elaborate antenna in the back garden he didn't have. It would have enabled him to chew the fat with postmasters in Kansas and trace his uncles who emigrated to the United States and were never heard from again; to visit the Library of Congress or see what the weather is like in San Francisco or check the photographs of his grandchildren that I post regularly on my Web server.

When my sister in Ireland discovered I was writing a book and that it opened with a story about Da, she went rummaging in her attic, and found a photograph of him that she found when clearing out our mother's house after her death. I have it before me now as I write. It's got a line down the middle where it's been folded for the best part of fifty years. But it's still a lovely photograph of a handsome young man with a high forehead and an intelligent, friendly face. It's the man who taught his rich friend Morse and dreamed of having his own rig. In the normal course of events I would frame it and put it on the wall. But this time I'm going to put it on the Web, where it belongs – and where you, dear reader, can see it too, if you wish.[14]

Notes

Chapter 1: Radio days

1 For a powerful evocation of the time, see George O'Brien, *The Village of Longing and Dancehall Days*, Penguin, 1987.
2 Nobody talked about frequency then.
3 Slang for amateur radio operators.
4 The ionosphere is really made up of three distinct layers which are usually labelled D, E and F. D extends from 50 to 90 km above the earth's surface; E extends from 90 to 160 km; and F extends from 160 km upwards. During the hours of darkness, the D layer effectively vanishes because its ionisation disappears in the absence of sunlight, so the layer which affects radio transmissions is 40 km higher than it is during the day.
5 The highest compliment he could pay to another Morse operator was that he had 'a lovely fist'. Later on, I discovered that this is what the operators who worked for Bletchley Park also used to say.
6 Available free from Real Networks at www.real.com.

Chapter 2: The digital beanstalk

1 The word 'bit' comes from 'BInary digiT'. It is the smallest indivisible piece of data that computers hold. Eight bits is a 'byte', and most computing operations involve handling thousands (kilo) or millions (mega) of bytes.
2 Actually it's a network of Digital Alpha workstations, but the network appears to the outside world as a single device.
3 For example, Kevin Kelly in *Out of Control: The New Biology of Machines*, Fourth Estate, 1994.
4 www.gip.org/gip1∅.htm

5 With one or two honourable exceptions. On 11 January 1999, for example, Alan Rusbridger, Editor of the *Guardian*, wrote a delightful piece in his paper on how he came to love the Net.
6 In contrast to his 1996 speech, when he did not refer to it once.
7 In the DRUDGE REPORT (www.drudgereport.com), 18 April 1997, admittedly a source hostile to Clinton.
8 Reuters report, 'French say "Oui!" to Web', published on NEWS.COM, 20 March 1998.

Chapter 3: **A terrible beauty?**

1 Henry Cabot Lodge (ed.), *The Education of Henry Adams: An Autobiography*, Constable, 1919, p. 380.
2 *Ibid.*, p. 381.
3 For up-to-date estimates see www.nua.ie.
4 The US National Science Foundation publishes some interesting statistics about the growth in the number of 'host' computers connected to the Net at ftp://nic.merit.edu/nsfnet/statistics/history.hosts.
5 Quotation from background information on http://www.altavista.com/av/content/about_our_story_2.htm
6 Gerry McGovern, 'New Thinking: Digital Pollution', 1 June 1998. McGovern runs Nua, a leading Internet consultancy company which publishes the Nua Internet Surveys. See archive at www.nua.ie.
7 *Daily Telegraph*, 12 December 1997.
8 *Time*, 26 June 1995.
9 *Time*, 24 July 1995.
10 www.human-nature.com/rmyoung/papers/pap108.html.
11 For more details see www.distributed.net/.
12 *Scientific American*, May 1997.
13 Bob Metcalfe was one of the graduate students who worked on the design of the ARPANET – the precursor to the Internet – and later went on to co-invent the Ethernet local area networking system.
14 George B. Dyson, *Darwin among the Machines: The Evolution of Global Intelligence*, Addison-Wesley, 1997.
15 *Ibid.*, p. 2.
16 Surveys conducted in late 1997 suggested that 21 per cent of US households had access to the World Wide Web, and that fully a quarter of these had signed up within the previous six months. And a study into 'The Buying Power of Black America' published in 1997 found that African American families spent more than twice as much per household for online services in 1996 as their white counterparts. The main reason cited for this rush to go online was the disappearance of distinctive Black

programming on mainstream US television and radio. See Steve Silberman, 'This Revolution Is Not Being Televised', *Wired News*, 11 September 1997.

17 Howard Rheingold, *The Virtual Community*, Secker & Warburg, 1994, p. 4.

18 *CMC Magazine*, February 1997, www.december.com/cmc/mag/1997/feb/weinon.html.

19 John Seabrook's *Deeper: A Two-Year Odyssey in Cyberspace*, Faber, 1997, provides a vivid description of the Well as experienced by a newcomer.

20 From his essay on the '24 Hours in Cyberspace' site – /www.cyber24.com/htm3/toc.htm?rescue.

21 Kelly, *op.cit.*, p. 9.

22 The conventional wisdom is that one 'Web year' equals seven chronological ones.

Chapter 4: **Origins**

1   In this case Roger Moore, Scott Forrest and John Monk.

2   E. M. Forster, *The Machine Stops and Other Stories*, ed. Rod Mengham, André Deutsch, 1997.

3   *The Times*, 1 November 1946, p. 2.

4   'The World's Latest Computer', *Economist*, 5 December 1998, p. 135.

5   Vannevar Bush, *Pieces of the Action*, Cassell, 1972.

6   Larry Owens, 'Vannevar Bush and the Differential Analyzer: The Text and Context of an Early Computer', *Technology and Culture*, vol. 27, no. 1 (January 1986), pp. 63–95. Reprinted in James M. Nyce and Paul Kahn (eds), *From Memex to Hypertext: Vannevar Bush and the Mind's Machine*, Academic Press, 1991.

7   Bush, *op. cit.*, p. 182.

8   The claim comes from an article by Leo Wiener about his educational philosophy published in the *American Magazine* of July 1911. Quoted in Steve J. Heims, *John von Neumann and Norbert Wiener: From Mathematics to the Technologies of Life and Death*, MIT Press, 1980, p. 4.

9   Heims, *op. cit.*, p. 1–2.

10 Norbert Wiener, *Ex-Prodigy: My Childhood and Youth*, Simon & Schuster, 1953, p. 67.

11 Norbert Wiener, *The Human Use of Human Beings: Cybernetics and Society*, Houghton Mifflin, 1950.

12 Wiener, *Ex-Prodigy*, p. 136.

13 Letter to Lucy Donnelly dated 19 October 1919 and quoted in I. Grattan-Guinness, 'The Russell Archives: Some New Light on Russell's Logicism', *Annals of Science*, 31 (1974), p. 406.

14 Ray Monk, *Bertrand Russell: The Spirit of Solitude*, Jonathan Cape, 1996, p. 326.

15 Norbert Wiener, *Cybernetics: or Control and Communication in the Animal and Machine*, Wiley, 1948, p. 7.

16 The five basic principles of the design were that: (1) it should be a numerical rather than an analog device; (2) it should be electronic rather than mechanical in operation; (3) it should use binary arithmetic; (4) the 'entire sequence of operations [should] be laid out on the machine itself so that there should be no human intervention from the time the data were entered'; and (5) the machine should 'contain an apparatus for the storage of data which should record them quickly, hold them firmly until erasure, read them quickly, erase them quickly, and then be immediately available for the storage of new material' (Wiener, *Cybernetics*, pp. 10–11).

17 Heims, *op. cit.*, p. 183.

18 It was also, according to Heims (*ibid.*, p. 335), a bestseller.

19 Wiener, *Cybernetics*, p. 9.

20 *Ibid.*

21 David Jerison, I. M. Singer and D. W. Stroock (eds), *The Legacy of Norbert Wiener: A Centennial Symposium*, American Mathematical Society, 1997, p. 19.

22 Though he later returned to MIT where he was Donner Professor of Science from 1958 to 1978.

23 Shannon's paper, 'A Mathematical Theory of Communication', was published in the *Bell System Journal* in 1948. It was republished a year later in book form together with an expository introduction by Warren Weaver (University of Illinois Press, 1949) – which is why one sometimes sees the theory referred to as the 'Shannon–Weaver' theory.

24 *Scientific American*, vol. 227, no. 3 (September 1972), p. 33.

25 Norbert Wiener, *God and Golem, Inc.: A Comment on Certain Points Where Cybernetics Impinges on Religion*, 1964, p. 71.

26 Wiener, *Cybernetics*, p. 37.

27 *Ibid.*, p. 38.

28 Heims, *op. cit.*, p. 333.

29 Quoted in *ibid.*, p. 379.

30 Wiener in *IEEE Annals of the History of Computing*. vol. 14, no. 2, 1992.

31 In 1945–6, the Navy allocated $875,000 to Forrester's project. By 1949 he was spending money at the rate of $100,000 *a month*! See Owens, *op. cit.*, p. 22.

32 Oral History Interview with Ken Olsen, Digital Equipment Corporation, by David Allison, Division of Information Technology & Society, National Museum of American History, Smithsonian Intitution, 28 and 29 September 1988. Transcript available online at www.si.edu/resource/tours/comphist/olsen.html.

33 Wesley A. Clark, 'The LINC Was Early and Small', in Adele Goldberg (ed.), *A History of Personal Workstations*, Addison-Wesley (ACM Press), 1988, p. 347.

34 Katie Hafner and Matthew Lyon, *Where Wizards Stay Up Late: The Origins of the Internet*, Simon & Schuster, 1998, p. 31.

35 For 'Programmed Data Processor'.

36 'An Interview with Wesley Clark', conducted by Judy O'Neill, 3.May 1990, New York, NY, transcript OH-195, Charles Babbage Institute, Center for the History of Information Processing, University of Minnesota, Minneapolis, Minnesota.

37 J. C. R. Licklider, 'Man–Computer Symbiosis', *IRE Transactions on Human Factors in Electronics*, vol. HFE-1, March 1960, pp. 4–11. Reprinted in *In Memoriam JCR Licklider*, Digital Equipment Corporation Systems Research Center, 7 August 1990.

38 'An Interview with J. C. R. Licklider' conducted by William Aspray and Arthur Norberg, 28 October 1988, Cambridge, MA, transcript OH-150, Charles Babbage Institute, Center for the History of Information Processing, University of Minnesota, Minneapolis, Minnesota.

39 *Ibid.*

40 Licklider, *op. cit.*

41 Martin Greenberger (ed.), *Management and the Future of the Computer*, MIT Press, 1962, p. 205.

42 The origins of time-sharing as an idea are confused by the fact that the term was first publicly used by Christopher Strachey, an Oxford computer scientist, in a different sense from that in which it was later (and universally) understood. At a conference in Paris in 1959, Strachey presented a paper entitled 'Time Sharing in Large Fast Computers' which was essentially about the sharing of central processor by multiple *programs* – as distinct from multiple *users*. Time-sharing in Strachey's sense was already an established technology by the time he gave his paper – it was used, for example, in SAGE.

43 John McCarthy, 'Reminiscences on the History of Time Sharing', 1983, available on his Stanford Website – www-formal.stanford.edu/jmc/.

44 The Symposium talks were printed in Martin Greenberger (ed.), *Computers and the World of the Future*, MIT Press, 1962.

45 Ronda Hauben, 'Cybernetics, Human–Computer Symbiosis and On-Line Communities: The Pioneering Vision and the Future of the Global Computer Network', available in various locations on the Web, for example www.columbia.edu/~hauben/netbook/ and also in Michael Hauben and Ronda Hauben, *Netizens: On the History and Impact of Usenet News and the Internet*, IEEE Computer Society, 1997.

46 Fernando Corbato and Robert Fano, 'Time-Sharing on Computers', in *Information – A Scientific American Book*, San Francisco, 1966, p. 76.

Chapter 5: Imps

1 'An Interview with Paul Baran' conducted by Judy O'Neill, 5 March 1990, Menlo Park, CA, transcript OH-182, Charles Babbage Institute, Center for the History of Information Processing, University of Minnesota, Minneapolis, Minnesota.

2 For a more detailed account of the origins and development of ARPA see Hafner and Lyon, *op. cit.*, pp. 11–42.

3 'An Interview with J. C. R. Licklider', *op. cit.*

4 *Ibid.*

5 *Ibid.*

6 *Ibid.*

7 Following initial funding by NASA under the aegis of Robert Taylor.

8 David Bennahum, 'The Intergalactic Network', http://memex.org/meme2-09.html.

9 Taylor is the unsung hero of modern computing, for in addition to conceiving and funding the ARPANET he went on to direct the Computer Science Laboratory at the Xerox Palo Alto Research Center, the lab which invented the core technologies of today's personal computers – the graphical user interface, the laser printer and the Ethernet networking system. The amazing story of that lab is told by Douglas Smith and Robert Alexander in their book *Fumbling the Future: How Xerox Invented, Then Ignored, the First Personal Computer*, Morrow, 1988; and by Michael Hiltzig's *Dealers of Lightning: XEROX PARC and the Dawn of the Computer Age*, Harper Business, 1999.

10 'An Interview with Robert A. Taylor' conducted by William Aspray, 28 February 1989, transcript OH-154, Charles Babbage Institute, DARPA/ITPO Oral History Collection, Center for the History of Information Processing, University of Minnesota, Minneapolis, Minnesota.

11 'An Interview with Ivan Sutherland' conducted by William Aspray, 1 May 1989, Pittsburgh, PA, transcript OH-171, Charles Babbage Institute, Center for the History of Information Processing, University of Minnesota, Minneapolis, Minnesota.

12 'An Interview with Robert A. Taylor', *op. cit.*

13 Robert A. Taylor, interviewed for *History of the Future*, a videotape commissioned by Bolt, Beranek and Newman to celebrate the twenty-fifth anniversary of the building of the ARPANET.

14 *Ibid.*

15 Hexadecimal is a number system which counts in sixteens rather than tens. Computer people like hexadecimal because two hexadecimal digits correspond to a single byte, thereby providing a convenient way of describing the contents of computer memory. Normal human beings find hexadecimal not only painful, but incomprehensible.

16 'An Interview with Lawrence G. Roberts' conducted by Arthur L. Norberg, 4 April 1989, transcript OH-159, Charles Babbage Institute, Center for the History of Information Processing, University of Minnesota, Minneapolis, Minnesota.

17 *Ibid.*

18 *Ibid.*

19 www.ziplink.net/~lroberts/. On his home page, Roberts describes himself as 'Founder of the Internet'.

20 Robert A. Taylor, e-mail to author, 5 December 1997.

21 Hafner and Lyon, *op. cit.*, p. 68.

22 Robert A. Taylor, e-mail to author, 5 December 1997.

23 See Thomas Marill and Lawrence G. Roberts, 'Toward a Cooperative Network of Time-Shared Computers', *Fall Joint Computer Conference of the American Federation of Information Processing Societies*, 1966.

24 Taylor, *History of the Future.*

25 Hafner and Lyon, *op. cit.*, print some reproductions of early Roberts sketches on pp. 48–50.

26 UCLA was chosen as the first site because ARPA had established a Network Measurement Center around Leonard Kleinrock there. Santa Barbara did advanced graphics research, as did Utah (in addition to researching night-vision for the military). Stanford Research Institute was chosen mainly because Doug Engelbart, the inventor of the computer 'mouse' (and of a great deal else besides) worked there. Engelbart re-enters our story later when we come to look at the precursors to the World Wide Web.

27 A seminal time-sharing project funded by ARPA.

28 'An Interview with Wesley Clark', *op. cit.*

29 *Ibid.*

30 Hafner and Lyon, *op. cit.*, p. 75.

31 Peter Salus, 'The Vogue of History', *Matrix News*, 6(7) July 1996.

Chapter 6: Hot potatoes

1 It later metamorphosed into the mainframe company, Univac, and then into Unisys.

2 'An Interview with Paul Baran,' *op. cit.*

3 *Ibid.* Mind you, there was a time (1943) when IBM Chairman Thomas Watson estimated the global demand for computers at five machines, none of which would be in business. (Charles Jonscher, *Wired Life: Who Are We in the Digital Age?*, Bantam Press, 1999, p. 32.)

4 Literally, metering at a distance.

5 The application of mathematical and statistical analysis to the planning

and execution of military operations. The name RAND comes from 'Research And Development'.

6 'An Interview with Paul Baran', *op. cit.*, p. 10.

7 *Ibid.*, p. 11.

8 J. M. Chester, 'Cost of a Hardened, Nationwide Buried Cable Network', RAND Corp. Memorandum RM-2627-PR, October 1960.

9 AM stands for 'audio modulated', AM stations are often referred to in common parlance as 'medium wave' or 'long wave' stations.

10 Dyson, *op. cit.*, p. 147.

11 Paul Baran, *Introduction to Distributed Communications Networks*, RAND Report RM-3420-PR, Figure 1.

12 'An Interview with Paul Baran', *op. cit.*, p. 18.

13 *Ibid.*, p. 23.

14 *Ibid.*, p. 24.

15 Paul Baran, 'On Distributed Communications', RAND Corp. Memorandum RM-3767-PR, August 1964.

16 Paul Baran, 'On Distributed Communications Networks', *IEEE Transactions on Communications Systems*, CS-12 (1964), pp. 1–9. The reports Baran and his colleagues produced during the survivable network project are available online as papers in the 'RAND Classic' series. See www.RAND.org/.

17 'An Interview with Paul Baran', *op. cit.*, p. 27. Two of the reports – one dealing with 'Weak spots and patches', the other with cryptography – were classified.

18 'An Interview with Paul Baran', *op. cit.*, p. 19.

19 Quoted in *ibid.*, p. 21.

20 Letter from F. R. Collbohm to Deputy Chief of Staff, Research and Development, United States Air Force, 30 August 1965, enclosing a paper entitled 'Limitations of Present Networks and the Improvements We Seek'.

21 'An Interview with Paul Baran', *op. cit.*, p. 35.

Chapter 7: **Hindsight**

1 University of Chicago Press, 1964.

2 Edward W. Constant II, *The Origins of the Turbojet Revolution*, Johns Hopkins University Press, 1980, p. 10.

3 *Ibid.*, p. 13.

4 *Ibid.*, p. 15.

5 AT&T corporate history, www.att.com.

6 See Ithiel de Sola Pool (ed.), *The Social Impact of the Telephone*, MIT Press, 1977.

7 Hafner and Lyon, *op. cit.*, p. 63.

8   *Ibid.*, p. 182.
9   *Ibid.*, p. 232.

Chapter 8: **Packet post**

1   David McCullough, *Truman*, Simon & Schuster, 1992, p. 564.
2   He noted Davies's recruitment with satisfaction, but his approval was tempered by annoyance at the way the young tyro pointed out some mistakes in the master's seminal paper, *Computable Numbers*.
3   'An Interview with Donald W. Davies' conducted by Martin Campbell-Kelly, 17 March 1986, National Physical Laboratory, UK, OH-189, Charles Babbage Institute, University of Minnesota, Minneapolis, Minnesota, p. 5.
4   For example, the Distributed Array Processor and a fast data-retrieval system called the Content Addressable File Store (CAFS).
5   John Bellamy, *Digital Telephony*, Wiley, 1982, p. 364.
6   Tom Standage, *The Victorian Internet: The Remarkable Story of the Telegraph and the Nineteenth Century's Online Pioneers*, Weidenfeld & Nicolson, 1998.
7   Bellamy, *op. cit.*, p. 366.
8   The name 'packet' was chosen with characteristic care, after checking that it had a cognate word in many languages.
9   Martin Campbell-Kelly, 'Data Communications at the National Physical Laboratory (1965–1975)', *Annals of the History of Computing*, vol. 9, no. 3–4, 1988, p. 226.
10 Donald Davies, 'Packet Switching: The History Lesson', *Guardian Online*, 21 August 1997.
11 Campbell-Kelly, *op. cit.*, p. 226.
12 Hafner and Lyon, *op. cit.*, p. 67. Davies was also the man who brought packet-switching to the attention of CCITT, the top-level standards body of the international telecommunications industry.
13 Campbell-Kelly, *op. cit.*, p. 226.
14 *Ibid.*, p. 229.
15 D. W. Davies, K. A. Barlett, R. A. Scantlebury and P. T. Wilkinson, 'A Digital Communication Network for Computers Giving Rapid Response at Remote Terminals', *ACM Symposium on Operating System Principles*, Gatlinburg, Tennessee, October 1967.
16 Lawrence G. Roberts, 'Multiple Computer Networks and Intercomputer Communication', *ACM Symposium, op. cit.*
17 *Ibid.*, p. 3.
18 This is significant because some versions of the ARPANET story have overlooked or underplayed the contributions of Baran and Davies.

Larry Roberts's personal website (www.ziplink.net/~lroberts/Internet-Chronology.html), for example, asserts that Leo Kleinrock wrote the 'first paper on packet-switching theory' in 1961 and followed it up with a 1964 book providing 'the network design and queuing theory necessary to build packet networks'. Inspection of the two publications, however, reveals no mention of *packet*-switching. Kleinrock was interested in *switching* networks generally, and his book provides a sublimely elegant mathematical analysis of the switching problem assuming variable message length, but it contains no mention of packets, although in an interview conducted by Judy O'Neill on 3 April 1990 for the Charles Babbage Institute (transcript OH-190), Kleinrock claims that the idea of a packet was 'implicit' in his 1964 book.

19 Hafner and Lyon, *op. cit.*, p. 76.
20 Campbell-Kelly, *op. cit.*, p. 232. The NPL researchers went on to build what was probably the world's first local area network – and certainly the first to provide a file server and direct service to simple terminals.
21 He eventually decided to use 50 kilobits per second leased lines.
22 'Invitation to tender' in European parlance.
23 'An Interview with Robert E. Kahn' conducted by Judy O'Neill, 24 April 1990, Reston, VA, transcript OH-192. Charles Babbage Institute, Center for the History of Information Processing, University of Minnesota, Minneapolis, Minnesota. Kahn was a cousin of Herman Kahn of the Hudson Institute, a noted (some would say notorious) 'futurologist' who wrote a bestselling book *On Thermonuclear War* and generally prided himself on 'thinking the unthinkable'.
24 Stephen D. Crocker, 'The Origins of RFCs', in J. Reynolds and J. Postel, *The Request for Comments Reference Guide* (RFC 1000), August 1987.
25 *Ibid.*
26 There is an archive of them at www.ietf.org/rfc.html.
27 Hafner and Lyon, *op. cit.*, p. 144.
28 Reynolds and Postel, *op. cit.*
29 Daniel C. Lynch, 'Historical Evolution', in Daniel C. Lynch and Marshall T. Rose (eds), *Internet System Handbook*, Addison-Wesley, 1993, p. 9.
30 Peter H. Salus, *Casting the Net: from ARPANET to Internet and Beyond . . .*, Addison-Wesley, 1995, p. 42. Campbell-Kelly, *op. cit.*, p. 231, thinks that the first use of the term 'protocol' in a data-communication context occurs in an NPL memorandum entitled *A Protocol for Use in the NPL Data Communications Network* written by Roger Scantlebury and Keith Bartlett in 1967.
31 Salus, *op. cit.*, p. 43.
32 Seabrook, *op. cit.*, pp. 80–1.

Chapter 9: **Where it's @**

1 Hafner and Lyon, *op. cit.*, p. 188.

2 Bob Taylor (private e-mail to author) says he was never surprised by it. But then Taylor had co-authored with Licklider a pathbreaking article, 'The Computer as a Communication Device', published in 1968. It was subsequently reprinted in *In Memoriam: J. C. R. Licklider 1915–1990*, Digital Systems Research Center, Palo Alto, 1990, pp. 21–31.

3 The IPTO's final report on the ARPANET project says: 'The largest single surprise of the ARPANET program has been the incredible popularity and success of network mail. There is little doubt that the techniques of network mail developed in connection with the ARPANET program are going to sweep the country and drastically change the techniques used for intercommunication in the public and private sectors.'

4 It was one of the IBM 360 series.

5 In 1999 Warner Brothers released *You've Got M@il!*, a romantic comedy written by Nora Ephron and starring Tom Hanks and Meg Ryan. Hanks and Ryan play rival bookstore owners in Manhattan who are bitter competitors in real life and lovers in Cyberspace. In fact, the film seems to have been a modern rewrite of a 1940 movies, *The Shop around the Corner*, with e-mail playing the role of snail mail.

6 Anne Olivier Bell (ed.), *The Diary of Virginia Woolf*, vol. 3: *1925–1930*, Penguin, 1982, p. 276.

7 Writing about the joys of e-mail in a newspaper supplement I once said that only a lunatic would dig out a postcard and stamp and go all the way to a post-box simply to say 'Hi! I was just thinking of you.' Two days later a postcard arrived from my most technophobic friend. 'Hi,' it read, 'I was just thinking of you.'

8 Robert M. Young, 'Sexuality and the Internet', www.human-nature.com/rmyoung/papers/pap108h.html.

9 *Ibid*.

10 Salus, *op. cit.*, p. 95. For some reason, Hafner and Lyon, *op. cit.*, p. 191, date it as 'one day in 1972', but this must be wrong because the RFC archive shows a flurry of discussions of a mail protocol in the summer and autumn of 1971.

11 Salus, *op. cit.*, p. 95 claims that 'internal mail was possible' by 17 July.

12 Quoted in a tribute to Tomlinson in *Forbes ASAP*, www.forbes.com/asap/98/1005/126.htm.

13 Though the definitive RFC630 was not released until 10 April 1974.

14 Eudora Lite is available from www.eudora.com. Pegasus comes from www.pmail.com.

15 Louisiana was a French possession which was sold to the fledgling United

States in 1803 for £3 million and admitted to the Union in 1812. In terms of natural resources it ranked second only to Texas.

Chapter 10: **Casting the Net**

1 Hafner and Lyon, *op. cit.*, p. 178.
2 Using the TELNET and FTP protocols hammered out by the students of the Network Working Group.
3 It was RFC089.
4 'Ah,' said a reader of an earlier draft, 'but two of the packets got lost in the system.' Yes, but remember that in a real interaction with the KPIX server any missing packets would be requested and retransmitted. One of the things that gives the Internet its power is that it was designed on the basis that the best (that is, guaranteed 100 per cent delivery of every packet first time) would be the enemy of the good (90.91 per cent success).
5 'An Interview with Robert E. Kahn', *op. cit.*
6 *Ibid.*
7 Hafner and Lyon, *op. cit.*, p. 178.
8 'An Interview with Robert E. Kahn', *op. cit.*
9 Quoted in Hafner and Lyon, *op. cit.*, p. 179.
10 Well, almost flawlessly: as reported earlier, the *one* time it crashed was when Bob Metcalfe was giving a demo to some suits from AT&T.
11 The random interval was important because if each station waited for a predetermined period before retransmitting the statistical chances of another collison turned out to be appreciable.
12 The legend of Cerf's trespass appears in several sources – for example, Hafner and Lyon, *op. cit.*, p. 139.
13 Vint Cerf, 'How the Internet Came To Be', in Bernard Aboba, ed., *The Online User's Encyclopedia*, Addison-Wesley, 1993.
14 *Ibid.*
15 Many of whom later became prominent in the Internet and communications world. They included Carl Sunshine, Richard Karp, Judy Estrin, Yogen Dalal, Darryl Rubin, Ron Crane and John Schoch. Bob Metcalfe also attended some of the discussions.
16 *IEEE Transactions on Communications*, COM-22, 5, pp. 637–48.
17 Hafner and Lyon, *op. cit.*, p. 227.
18 For 'Defense Advanced Research Projects Agency', Putting the 'D' up front was probably a sop to Congressional critics who wondered what a military funding body was doing sponsoring way-out ideas like electronic mail.
19 At one point it even offered it to AT&T. But the telephone company's suspicions of packet-switching persisted and the offer was refused!

20 Internet Experiment Notes were the INWG's equivalent of the ARPANET's RFCs.

21 Cerf, *op. cit.*

22 As with the RFCs, there are IEN archives on various servers worldwide – see for example sunsite.doc.ic.ac.uk/computing/internet/rfc/ien/ien-index.txt.145.

23 Good popular expositions of PARC's contributions can be found in Stephen Levy, *Insanely Great: The Life and Times of Macintosh*, Viking, 1994, and Robert X. Cringely, *Accidental Empires: How the Boys of Silicon Valley Make their Millions, Battle Foreign Competition, and Still Can't Get a Date*, Viking, 1992.

24 John Shoch, private communication, 29 October 1995.

25 Cerf, *op. cit.*

Chapter 11: **The poor man's ARPANET**

1 Dennis Ritchie, 'The Development of the C Language', ACM *Second History of Programming Languages Conference*, Cambridge, MA, April 1993, p. 1.

2 Ned Pierce, 'Putting Unix in Perspective', interview with Victor Vyssotsky, *Unix Review*, January 1985, pp. 60–2.

3 The capitalisation of the name has consumed acres of paper in its time. Some people think it should be called 'Unix'. I've opted for the capitalised version.

4 G. Pascal Zachary, *Show-Stopper!: The Breakneck Race to Create Windows NT and the Next Generation at Microsoft*, Little Brown, 1994, p. 266. The comparison is not entirely fair because NT does more than the UNIX kernel. But even when the other elements needed to provide similar functionality in UNIX are added in, it is still orders of magnitude less bloated than NT.

5 M. D. McIlroy, E. N. Pinson and B. A. Tague 'Unix Time-Sharing System Forward', *Bell System Technical Journal*, vol. 57, no. 6, part 2 (July–August 1978), p. 1902.

6 For example, the kernel does not support file access methods, file disposition or file formats. Neither does it concern itself with print spooling, command language, assignment of logical file names and so on.

7 Pierce, *op. cit.*

8 Like much else in this story, the nomenclature was accidental. Ritchie's search for a new language led him to revise a language called B in honour of its own parent BCPL, a language which originated at Cambridge (UK). In the circumstances, the next letter of the alphabet probably seemed an obvious choice. Hackers are like that.

9 The example is taken from the nearest thing C programmers have to a bible

– Brian W. Kernighan and Dennis M. Ritchie, *The C Programming Language*, Prentice-Hall, 1978, p. 6.

10 *Unix Review*, October 1985, p. 51.

11 John Stoneback, 'The Collegiate Community', *Unix Review*, October 1985, p. 27.

12 One source (Henry Edward Hardy, 'The History of the Net', Master's thesis, School of Communications, Grand Valley State University, 1993) credits the revised version to Lesk, David Notowitz and Greg Chesson.

13 For those familiar with the IMB PC a UNIX shell script is akin to an MS-DOS batch file.

14 Quoted in Hauben and Hauben, *op. cit.*, p. 40.

15 System managers at some industrial research labs – for example, Bell Labs and DEC – helped by subsidising some of the telecommunications costs of operating Usenet.

16 Rheingold claims that this happened sometime in 1981. Rheingold, *op. cit.*, p. 121.

17 Collected by Gene Spafford and David C. Lawrence and published in the Usenet History Archives. The figures for 1993 come from Lawrence's update of Usenet statistics for that year. See Hauben and Hauben, *op. cit.*, p. 47.

18 For an overview see www.dejanews.com, a Website which provides browser access to Usenet. An expert who gave evidence in the case against the Communications Decency Act claimed there were upwards of 15,000 news groups active. But because Usenet is not a centralised system, and news groups are being added every week, nobody knows for sure how many groups there are.

19 Hardy, *op. cit.*

20 Usenet used the period punctuation to denote hierarchical paths. Thus rec.sex means a news group called 'sex' in the hierarchy devoted to recreation.

21 Hardy, *op. cit.* Rmgroup is a News command to remove a group from the system.

22 Quoted in *ibid.*

23 Rheingold, *op. cit.*, p. 6.

Chapter 12: **The Great Unwashed**

1 Slang for Assembly Language, the low-level programming language which is just one remove from the code which actually drives the hardware.

2 This reconstruction relies heavily on the much fuller account given in Rheingold, *op. cit.*

3 *Ibid.*, p. 131.

4 *Ibid.*, p. 132.
5 It was one hell of a mongrel – a machine with a Motorola 68000 processor (the chip which powered the original Apple Macintosh) running a special version of MS-DOS which Jennings had written for it.
6 Rheingold, *op. cit.*, p. 137.
7 *Ibid.*, p. 139.
8 Randy Bush, 'A History of Fidonet', www.well.com/user/vertigo/history.html. The word 'Fidonet', incidentally, is a trademark of Tom Jennings.
9 For instructions see the Fidonet Website – www.fidonet.org/.
10 The full text can be found at www2.epic.org/cda/cda_dc_opinion.html.
11 The Supreme Court judgment is reproduced on various legal sites – for example, supct.law.cornell.edu/supct/html/96–511.ZO.html.
12 Not to mention synchronous communication via mechanisms like Internet Relay Chat – a facility through which you can contribute in 'real time' to a global conversation simply by joining in and typing whatever it is you want to contribute.

Chapter 13: **The gift economy**

1 See Stephen R. Graubard (ed.), *The Artificial Intelligence Debate: False Starts, Real Foundations*, MIT Press, 1989, for some thoughtful reflections on the crisis.
2 He has written two remarkable books setting out his views – *Mindstorms: Children, Computers and Powerful Ideas* (Basic Books, 1980) and *The Children's Machine* (Prentice-Hall, 1993).
3 Andrew Leonard, profile of Richard Stallman, *Salon* (www.salonmagazine.com), 31 August 1998.
4 Charles C. Mann, 'Programs to the People', *Technology Review*, January/February 1999.
5 Leonard, *op. cit.*
6 www.gnu.ai.mit.edu/.
7 www.gnu.ai.mit.edu/philosophy/philosophy.html. Software that is free in the Stallman sense should not be confused with software that is given away. As I write, for example, Microsoft does not charge for its Internet Explorer suite of browser programs. But these come to me not in their original source code, but as *compiled* code which is essentially a meaningless, enormously long, sequence of ones and zeroes. There is no way I could even begin to modify one of these programs, so they violate the first of Stallman's criteria.
8 After nine years AT&T sold UNIX to the networking company Novell, which two years later, in 1995, sold it on to the Santa Cruz Operation

(SCO). Nowadays variants of UNIX are available from companies such as SCO, IBM, Digital, HP and Sun.

9   Leonard, *op. cit.*

10 Recursion is a programming trick which involves defining a function or a procedure in terms of itself.

11 Quoted in Mann, *op. cit.*

12 *Ibid.*

13 Andrew S. Tanenbaum, *Operating Systems: Design and Implementation*, Prentice-Hall, 1987. MINIX was a marvel of miniaturisation, designed so that it could be run on a dual-floppy IBM PC.

14 Although it's been overshadowed by what Torvalds built on it, MINIX is still going strong. The source code is copyrighted, but can be downloaded from www.cs.vu.nl/~ast/minix.html so long as you agree to the licensing requirements.

15 Glyn Moody, 'The Greatest OS that (N)ever Was', *Wired*, August 1997.

16 *Ibid.*

17 'bash' (which stands for 'Bourne Again Shell', a typical hacker's joke because it's a revision of the original Bourne shell) is the most common UNIX shell – the basic command-line interface to the operating system. 'gcc' refers to the GNU C-compiler.

18 This copy comes from theory.ms.ornl.gov/~xgz/linus_announce.

19 Because a complete Linux implementation blends a descendant of Torvald's original kernel with software utilities from Stallman's GNU project, it ought strictly to be called GNU–Linux. But this subtlety is lost in most media coverage of the Linux phenomenon.

20 Josh McHugh, 'Linux: The Making of a Global Hack', *Forbes*, 10 August 1998. Online version at www.forbes.com/forbes/98/0810/6203094a.htm.

21 In 1998, Linux was paid the ultimate backhanded compliment, for an internal Microsoft memorandum leaked on the Net revealed that the company was beginning to see Torvalds's creation as a serious potential competitor for its key strategic product – Windows NT.

22 See www.opensource.org/ for more information.

23 www.redhat.com.

24 McHugh, *op. cit.*

25 Eric S. Raymond 'Homesteading the Noosphere', www.tuxedo.org/~esr/ writings/homesteading/.

Chapter 14: **Web dreams**

1   The name came from 'visual calculator'.

2   Cringely, *op. cit.*, p. 67.

3   For a detailed account of Bush's involvement see Robert Buderi, *The*

Invention that Changed the World: The Story of Radar from War to Peace, Little
Brown, 1997.

4 *Ibid.*, p. 34.
5 James M. Nyce and Paul Kahn (eds), *From Memex to Hypertext: Vannevar
   Bush and the Mind's Machine*, Academic Press, 1991, p. 40.
6 Vannebar Bush, 'The Inscrutable 'Thirties', *Technology Review*, 35(4), pp.
   123–7.
7 Vannevar Bush, 'Memorandum Regarding Memex', in Nyce and Kahn, *op.
   cit.*, p. 81.
8 *Ibid.*
9 Indeed one of the ironies of Bush's life is that his great Differential Analyzer
   was rendered obsolete by the advent of digital computation.
10 Meaning a Xerox-type process. A lot of Bush's earlier research had been
   into this kind of photographic technology.
11 'The Augmented Knowledge Workshop: Participants' Discussion', in Goldberg (ed.), *op. cit.*, p. 235.
12 His way of implementing Bush's vision of racks of cathode-ray tubes each
   giving a different view of a single file, or views of different but related files.
13 The videotape is now on display at the Smithsonian Museum Exhibit on
   'The Information Age'.
14 Levy, *op. cit.*, p. 42.
15 *Ibid.*
16 The demonstration was promoted and funded by ARPA, at Bob Taylor's
   instigation.
17 In 1997 he was awarded the annual $500,000 Lemelson–MIT prize for the
   invention of the mouse.
18 Douglas C. Engelbart, 'Dreaming of the Future', *Byte*, September 1995.
19 This is even more plausible for Joyce's *Finnegans Wake*. I've seen a claim (in
   www.mcs.net/~jorn/html/jj.html) that 'exhaustive hypertext annotations
   of the puns and allusions in a single paragraph can amount to 100kb'.
   It goes on to reference a study of Chapter Four, Paragraph One
   (www.mcs.net/~jorn/html/jj/fwdigest.html) collected on the FWAKE-L
   mailing list in 1991.
20 Computer-industry lingo for exciting software which fails to appear.
21 Gary Wolf, 'The Curse of Xanadu', *Wired*, 3 June 1995.
22 *Ibid.*
23 *Ibid.*
24 Ted Nelson, 'All for One and One for All', *Hypertext '87 Proceedings*,
   November 1987.
25 A pixel is an element of a graphical display representing the brightness and
   possibly colour of a part of an image. The term derives from 'picture
   element'.
26 Levy, *op. cit.*, p. 187.

27 'Bill Atkinson: The Genius behind Hypercard', interview with *Quick Connect*, November 1987. The interview is archived at: http://www.savetz.-com/ku/ku/quick_genius_behind_hypercard_bill_atkinson_the_november 1987.html.

28 See Alan Kay, 'The Dynabook – Past, Present and Future', in Goldberg (ed.), *op. cit.*, pp. 253–63.

29 Levy, *op. cit.*, p. 241.

30 Atkinson, *Quick Connect* interview.

31 *Ibid.*

32 Levy, *op. cit.*, p. 247.

33 The English equivalent is Meccano.

34 Atkinson, *Quick Connect* interview.

35 *Ibid.*

Chapter 15: **Liftoff**

1 Charles Simonyi Professor of the Public Understanding of Science, Oxford University, writing in the *Guardian*, 11 November 1996.

2 www.w3c.org.

3 Tim Berners-Lee, 'Information Management: A Proposal', CERN internal report, May 1990, p. 9. Available online at www.w3.org/History/1989/proposal.html.

4 *Ibid.*

5 *Ibid.*

6 *Ibid.*

7 *Ibid.*

8 *Ibid.*

9 See Nicole Yankelovich, *et al.*, 'Issues in Designing a Hypermedia Document System', in Ambron and Hooper (eds), *Interactive Multimedia: Visions of Multimedia for Developers, Educators, and Information Providers*, Microsoft Press, 1988, pp. 33–85.

10 FTP was – and remains – the standard way to transfer programs and data across the Net.

11 For example, GOPHER and WAIS. GOPHER is a menu-driven system invented at the University of Minnesota which enabled users to browse the Internet looking for resources. WAIS ('Wide Area Information Servers') is a system for searching for indexed resources on the Net.

12 Robert H. Reid, *Architects of the Web: 1,000 Days that Built the Future of Business*, Wiley, 1997, p. 5.

13 Frank G. Halasz, 'NoteCards: A Multimedia Idea-Processing Environment', in Ambron and Hooper, *op. cit.*, pp. 105–9.

14 Available as a free download from www.aolpress.com.

15 Berners-Lee summarised it as 'a generic stateless object-oriented protocol'.
16 George Gilder, 'The Coming Software Shift', *Forbes*, 28 August 1995.
17 Gary Wolf, 'The (Second Phase of the) Revolution Has Begun', www.hotwired.com/wired/2.10/features/mosaic/html.
18 Marc Andreessen, interview with Thom Stark, www.dnai.com/~thomst/marca.html.
19 Gilder, *op. cit.*
20 Graphics Interchange Format – a standard for exchanging graphics files originally developed by CompuServe.
21 Programmers call this 'porting' (from 'transporting').
22 Reid, *op. cit.*, p. 9.
23 By this stage, the US National Science Foundation was paying for a large slice of the Internet's infrastructure.
24 Gilder, *op. cit.*
25 Reid, *op. cit.*, p. 12.
26 *Ibid.*
27 See Donald A. Norman, *The Invisible Computer*, MIT Press, 1998, p. 7.
28 Reid, *op. cit.*, pp. 17–18.
29 *Ibid.*, p. 17.
30 *Ibid*, p. 20.
31 It was originally called Mosaic Communications.
32 Technically, it was free to students and educators. Everyone else was supposed to pay $39. But even they got Beta (that is, pre-release) versions for nothing.
33 Joshua Quittner and Michelle Slatalla, *Speeding the Net: The Inside Story of Netscape and How It Challenged Microsoft*, Atlantic Monthly Press, 1998, p. 210.
34 James Collins, 'High Stakes Winners', *Time*, 26 February 1996, p. 44.

Chapter 16: **Home sweet home**

1 James Wallace, *Overdrive: Bill Gates and the Race to Control Cyberspace*, Wiley, 1997, p. 105.
2 Quittner and Slatalla, *op. cit.*, p. 190. Their account of the Sinofsky trip and its aftermath is corroborated in Wallace, *op. cit.*, p. 148.
3 Nathan Myhrvold, Gates's technology guru, was once heard to joke that 'software is like a gas – it expands to fill the space available'.
4 Quittner and Slatalla, *op. cit.*, p. 283.
5 Eric S. Raymond, 'The Cathedral and the Bazaar', original at sagan.earthspace.net/esr/writings/cathedral-bazaar/. There is an HTML version on the RedHat site – www.redhat.com/redhat/cathedral-bazaar/cathedral-bazaar.html.

6 Frederick P. Brooks, *The Mythical Man-Month: Essays in Software Engineering*, Addison-Wesley, 1975.

7 *Ibid.*, p. 42.

8 Eric S. Raymond, interview by Andrew Leonard, *Salon* (www.salonmagazine.com/), 14 April 1998. Bind is the program which 'resolves' domain names – that is, translates addresses like www.disney.com into a numerical IP address. PERL (acronym for 'Practical Extraction and Support Language') is a programming language designed for scanning text files and extracting information for them. It is very widely used on the Net – indeed someone once described it as 'the duct tape of the Web'. Sendmail is the program which handles upwards of 80 per cent of the world's e-mail. Apache – as we saw in Chapter 13 – is the Web's favourite server program.

9 Helped, of course, by the fact that it is free, or relatively inexpensive when purchased as part of a packaged distribution which includes installers and other supporting applications.

Epilogue: the wisdom of the Net

1 If you're really interested, the yacht drifts north-east at 14.14 knots.

2 Proverbs, 29:18.

3 Levy, *op. cit.*, p. 52.

4 Cringely, *op. cit.*

5 Jerome Bruner, 'The Conditions of Creativity', in his *On Knowing: Essays for the Left Hand*, Harvard University Press, 1962, p. 18.

6 E. P. Thompson, *The Making of the English Working Class*, Penguin, 1963.

7 For an elegant exposition of what the history of the Net looks like from this viewpoint see Brian Winston, *Media, Technology and Society: A History from the Telegraph to the Internet*, Routledge, 1998.

8 Lawrence Lessig, 'Open Code and Open Societies: The Values of Internet Governance', 1999 Sibley Lecture, University of Georgia, Athens, Georgia, 16 February, 1999.

9 *Ibid.*

10 Raymond, 'The Cathedral and the Bazaar'.

11 Dave Clark, Internet Engineering Task Force (IETF), 1992.

12 Lessig, *op. cit.*

13 *Ibid.*

14 www.briefhistory.com/da/.

A note on sources

There is already an extensive literature on the history of the Internet – much of it held on the network itself. There are also some excellent books on the subject, though none with the scope that I have attempted in this volume. Each tends to illuminate only a part of the story, leaving the remainder to be filled in from other sources.

In 1996, for example, Katie Hafner and Matthew Lyon published the first genuinely popular account of the development of the ARPANET and the early days of the Internet – *Where Wizards Stay Up Late: The Origins of the Internet* (Simon & Schuster, 1996). This is a fine book but its scope is inevitably limited because it ends with the party to celebrate the twenty-fifth anniversary of the network and thus leaves out the World Wide Web entirely. A year earlier, Peter Salus, a scientist with a good knowledge of the Net, had published *Casting the Net: From ARPANET to INTERNET and Beyond . . .* (Addison-Wesley, 1995), a useful but idiosyncratic and at times highly technical account of the story. And although Salus's narrative takes him up to 1994, he mentions the Web only in passing, even though by that time it was three years old.

For their part, histories of the World Wide Web tend to take the network on which it rides more or less as given. For example Robert Reid's *Architects of the Web: 1,000 Days that Built the Future of Business* (Wiley, 1997), a collection of profiles of the movers and shakers of the Web, includes only a cursory nod in the direction of ARPANET and doesn't really take up the story of the Web until three years after Tim Berners-Lee invented the thing at CERN. And *Speeding the Net: The Inside Story of*

Netscape and How It Challenged Microsoft by Joshua Quittner and Michelle
Slatalla (Atlantic Monthly Press, 1998), suffers from the same problem – it
opens with the founding of Netscape in 1994, and thus also ignores the
early period of the Web's development, as well as its extensive pre-
history.

There are a number of other books which are indispensable for anyone
seeking to understand the evolution of the Net. For years, a prime source
was the Netizens Website on the origins and significance of Usenet,
which was created and maintained by Michael and Ronda Hauben at
Columbia University (www.columbia.edu/~hauben/netbook/). In 1997
the Haubens bowed to the inevitable and issued a hard copy of their
marvellous site as *Netizens: On the History and Impact of Usenet and the
Internet* (IEEE Computer Society, 1997).

The origins and early years of networking and conferencing have also
been memorably and sensitively chronicled by Howard Rheingold in two
splendid books – *Tools for Thought* (Simon & Schuster, 1985) and *The
Virtual Community: Finding Connection in a Computerized World* (Secker &
Warburg, 1994). When *Tools for Thought* went out of print and its
publisher inexplicably declined to reissue it, Rheingold asked for the
rights to revert to him and published it on the Net, where it still remains,
freely available to all at www.rheingold.com. Visit it, and then raise a
glass to the short-sightedness of publishers.

As far as the history of computing is concerned, the reader is already
spoiled for choice. My own favourite is *Computer: A History of the
Information Machine* by Martin Campbell-Kelly and William Aspray
(Basic Books, 1996), a good general history of the digital computer from
earliest times to the Internet. And the best study of the evolution of
communications technologies in their socio-economic context is un-
doubtedly Brian Winston's *Media Technology and Society: A History from the
Telegraph to the Internet* (Routledge, 1998), though his reliance on a model
drawn from Saussurian linguistics might make some engineers choke on
their muesli.

Since much of the essential documentation – and almost all of the
primary sources – for the history of the Net is online, it would be
ridiculous to attempt to compile a definitive bibliography via the
venerable process of squirting vegetable dyes on to processed wood
pulp. And while I have tried to give full references to online sources in

the Notes, I am also conscious of the speed with which things change on the Net, and of the likelihood that the reader will sometimes be confronted by that bane of the Web-searcher's existence: the 'Error 404 – page not found' message.

If you do run into this virtual wall, then do not despair, because I maintain an up-to-date set of references where you can always get at them – on the Net. And if you find other sources that you think I, or other readers, might find useful, do please let me know. See you at www.brief history.com.

Glossary

ALOHA: The first **packet**-based* radio network, built by Norm Abramson and his colleagues at the University of Hawaii.

Alpha: A very early version of a program, generally the first version considered stable enough to be released to trusted collaborators and testers. Alpha versions are usually full of **bugs** and are often incomplete, though generally the entire framework of the envisaged program is sketched out.

AltaVista: An Internet search engine created and operated by Digital Equipment Corporation. Now owned by Compaq and ripe for flotation.

Analog: (analogue in UK English): Literally a model of some real-world phenomenon. In communications it means an electrical **signal** which varies smoothly and continuously over time in contrast to **digital**, which describes a signal encoded as a stream of ones and zeros, i.e. **bits**.

APACHE: The world's most popular Web-sever program. Developed within the **Open Source** tradition, Apache powers around 50 per cent of all the **servers** on the Web.

ARPA: The Advanced Research Projects Agency of the US Department of Defense. Established under President Eisenhower to oversee all advanced federally supported scientific and technological research.

ARPANET: The original **packet-switched** network, funded by the Defense Advanced Research Projects Agency and first operational in late 1969.

---

*Terms in boldface type are defined elsewhere in the Glossary.

Assembler: A program for translating programs written in **Assembly Language** into **machine code**. Analogous to a **compiler** for a higher-level programming language.

Assembly Language: A language for expressing **machine code** instructions in readable terms.

Bandwidth: A measure of the capacity of a communications channel to pass information down it. Originally measured in terms of the range of frequencies a channel could carry, it is increasingly measured in bits per second. Fibre-optic cable has incredible bandwidth.

BBS: See **Bulletin Board**.

Bell Labs: The R&D organisation of AT&T. Famous for inventing, among other things, the transistor, the laser and the **UNIX** operating system.

Beta: The penultimate version of a program, generally sent to volunteer users for testing prior to release. Beta versions are generally full working versions, but with various limitations. The program may still crash in quite ordinary circumstances, for example, and may take up more memory and processor time than the optimised first release. Custom and practice in the industry is to label Beta versions up to 0.9 with version 1.0 being the first official release.

Binary: The number system used by digital computers. Literally means 'base two'. There are only two numbers in the binary system – 0 and 1. In everyday life we use the denary (base 10) system. The number 13 in denary (one ten plus three units) becomes 1101 in binary (one eight [two cubed] plus one four [two squared] plus zero twos plus one unit).

Bit: Binary Digit – the smallest unit of data a computer can handle. A one or a zero.

Broadband: High **bandwidth** as in 'broadband network'. Sometimes also called wideband.

Browser: A program running as a **client** on a computer and providing a **virtual** window on to the **World Wide Web**.

Browsing: Using a browser to access documents on the Web. Often called **surfing**, though browsing has less frivolous connotations.

Bug: An error in a computer program.

Bulletin Board: Often abbreviated to BBS. A computer system used for posting messages which can be read by other subscribers. A kind of **virtual** notice board.

BYTE: Eight bits. Also the title of a legendary computer magazine.

C: High-level programming language developed at **Bell Labs** in association with **UNIX**. C or its derivative C++ is still the preferred language of applications programmers because it combines the advantages of a high-level language with facilities for accessing the machine hardware directly, much as Assembler does.

CDA: Communications Decency Act, the first serious attempt to regulate the content of Internet communications, passed by the US Congress in 1996, declared unconstitutional by the Supreme Court just over a year later.

Central Processing Unit: Usually abbreviated to CPU. The microprocessor (or **chip**) which does the actual data processing.

Chip: Colloquial term for an integrated circuit etched on to a wafer of **silicon**. Usually implies a microprocessor (i.e. a **CPU**) or **RAM**.

Circuit switching: A switching system, originally associated with the **analog** telephone system, whereby a physical connection was established between a transmitter and receiver, by means either of manual switchboards or of automatic electro-mechanical or electronic switches.

Client: A computer program, generally running on a networked computer, which provides services (e-mail, Web browsing, etc.) to its user by interacting with another computer on the network called a **server**.

Compiler: A program which translates instructions written in high-level programming languages like C or BASIC into machine code which the computer's CPU can execute. See **source code**.

Copyleft: A licensing system invented by Richard Stallman and the Free Software Foundation which grants to users of a program the right to alter its **source code** provided they pass on the right to alter the revised code under the same terms. Copyleft is the basic licensing system which underpins the **Open Source** movement.

CPU: See **Central Processing Unit**.

Cracker: Someone who gains unauthorised access to computer networks. Usually confused by the mass media with **hacker**.

Cybernetics: The science of control systems, mainly based on the idea of **feedback**. The term was coined by Norbert Wiener in the later 1940s from the Greek word for 'steersman'. Cybernetic systems are, in effect, self-steering ones. See also **servomechanism**.

Cyberspace: The world behind your screen. The term was coined by writer William Gibson in his 1984 science-fiction novel, *Neuromancer*.

DARPA: See **ARPA**.

DEC: Digital Equipment Corporation founded by Ken Olsen, an MIT researcher; in the 1970s and 1980s DEC was a leading manufacturer of minicomputers. The company was overtaken by the personal computer revolution and eventually bought by Compaq, the original manufacturer of IBM PC clones.

Digerati: Term coined to describe the smug elite of the IT revolution. Often applied to those who write for *Wired* magazine.

Digital: Literally 'of digits', but usually used to denote anything involving the use of **bits**. Contrasts with **analog**.

DOS: Literally 'disk operating system' but nowadays universally associated with the operating system which drove the first generations of IBM-PC-compatible computers. See **MS-DOS**.

Encryption: Encoding digital data so that they cannot be read by unauthorised third parties. A very hot topic because of the increasing volume of sensitive messages (e.g. credit card numbers) being transmitted across the Net.

Ethernet: A type of Local Area Network or **LAN** invented by Bob Metcalfe and others at XEROX **PARC** in the early 1970s and inspired in part by the **ALOHA** packet-radio network.

Explorer: Generic name for Microsoft's family of Web **browsers**.

Fidonet: A global store-and-forward network of computers created by and for personal computer users (as distinct from users of permanently networked systems common in universities and research laboratories).

Feedback: The relaying of information about the state of a system to its controller. The underlying requirement for automatic control. See also **Cybernetics**.

Fibre Optics: The science and technology of transmitting light through glass wire. A glass fibre is a fine glass filament which transmits light just as copper wire conducts electricity. Increasingly important in communications because of its phenomenal **bandwidth**. It's been estimated that one single strand of fibre-optic cable the thickness of a human hair could carry all the telephone conversations in the US simultaneously.

Floppy disk: A magnetically coated disk capable of storing data. The disk is flexible – hence the name – but is generally encased in a hard plastic shell for protection. Nowadays often alled a 'floppy'. Compare with **hard disk**.

FTP: File Transport Protocol. The second oldest network protocol, first agreed for ARPANET in 1970 and still an essential part of the Net. FTP is a set of technical conventions for the secure transfer of files across the network from one computer to another. Sometimes used as a verb – e.g. 'Can you FTP me the file?'

Geek: Someone who is more interested in (computer) technology than in the ordinary business of living. Often used semi-ironically by computer people. See also **nerd**.

GIF: Graphics Information File – a widely used standard, originally developed by CompuServe, for storing pictures. Most images on Web pages (except those authored by members of the **Open Source** movement) tend to be GIFs. See also **JPEG**.

Gopher: A suite of programs popular in pre-**WWW** days which helped users find information on the Net. Gopher can find information by subject category and was named after the mascot of the University of Minnesota, where it originated.

Graphical User Interface: Usually called a **GUI**. What you see when you boot up a Windows or a Macintosh computer – i.e. a user interface based on pictures or icons.

GUI: See **Graphical User Interface**.

Hacker: In tabloid demonology, an evil genius who specialises in breaking military and banking networks in order to wreak havoc. In the computer business and especially in the **Open Source** movement, the term 'hacker' is a compliment. It means a gifted programmer. Real hackers use the term **cracker** to describe the aforementioned tabloid demons.

Handshake: What goes on between two machines when they are establishing a method of communicating with one another. The noise a modem makes when it's connecting to an ISP's modem is the sound of the two devices negotiating a digital handshake.

Hard disk: A magnetically coated disk capable of holding magnetically encoded data. The name comes from the fact that the disk is rigid. Unlike a floppy, it is also perpetually spinning – at high speeds – and has much faster performance in recording and retrieving data.

Hardware: The most tangible, physical part of a computer system. The stuff that crushes your finger if it falls on it.

Host: Generically, a big computer connected to a network. In the ARPANET, the hosts were the big mainframe machines at each site.

HTML: Hypertext Mark-up Language. A subset of SGML (Standard Generalised Mark-up Language) developed by Tim Berners-Lee in 1989–90 at CERN. Basically a way of annotating Web pages so that a **browser** can display them in a manner approximating to their authors' intentions.

HTTP: Hypertext Transport Protocol – the technical conventions which govern transactions between **WWW client** programs and **servers** which dispense Web pages upon request.

HyperCard: A **hypertext** program developed by Bill Atkinson for the Apple Macintosh. Based on the idea of having infinitely extensible stacks of virtual index cards which could be hot-linked to one another by simple programming.

Hyperlink: A link within a Web page which, when activated, brings up another part of the same page, or another page located elsewhere either on the user's hard disk or on the Web.

Hypermedia: As **hypertext**, but mixing media other than printed text.

Hypertext: Literally, non-linear writing. Text that need not be read from beginning to end but browsed non-sequentially by following internal links. The underlying metaphor for the **World Wide Web**.

IMG tag: One of the features added to **HTML** by the authors of **Mosaic** which made it easy to include graphic images in a Web page.

IMP: Interface Message Processor – the Honeywell minicomputer which served as the nodes of the **ARPANET**.

Internet: The global network of computer networks.

Internet Relay Chat: A system which enables an **Internet** user who has installed the necessary client software to participate in live 'real-time' discussions with other Internet users in virtual 'chat rooms'.

IPTO: Information Processing Techniques Office. The department of ARPA which conceived and funded the ARPANET. Originally called the Command and Control Division, the department was renamed by J.C.R. Licklider while he was its Director.

JPEG: Acronym for Joint Photographic Expert Group, more commonly used as shorthand for a special way of compressing still images. Many images in Web pages are called JPEGs for this reason. JPEG is the image format used by the Open Source movement because the alternative **GIF** format involves the use of copyrighted technology.

LAN: Local area network. Contrasts with **WAN**.

LINUX: Pronounced LINN-ux. A clone of the **UNIX** operating system inspired by **MINIX**, created by a Finnish student named Linus Torvalds and distributed free on **Open Source** licensing terms. Linux is not a commercial product, yet runs a significant proportion of the world's **Internet** and local area network **servers** because of its remarkable stability and reliability.

Machine Code: The operation code of a particular **CPU**.

Memex: Vannevar Bush's term for a machine which could record, store and retrieve information quickly and efficiently and allow the user to create 'associative trails' through any amount of documentation. An early model for the Web in fact.

MINIX: A tiny functional clone of **UNIX** developed by Andrew S. Tanenbaum for teaching purposes.

Modem: Literally, a 'modulator de-modulator'. A device which translates digital signals into audio tones capable of being transmitted down an ordinary telephone line, and translates received audio tones back into streams of ones and zeros.

Mosaic: The first major graphics-based browser, written by Marc Andreessen and Eric Bina at NCSA in 1993.

MS-DOS: The original Microsoft operating system for IBM-PC-compatible computers.

Navigator: Netscape's Web **browser**. The first version was launched in 1994.

Nerd: Someone, generally young, generally male, who is more interested in computers than socialising. Traditionally a term of abuse until the mass media began to appreciate how rich and powerful nerds like Bill Gates were becoming. See also **geek**.

Netizen: A citizen of **Cyberspace**; someone who uses the Net a lot and subscribes to its original collective, co-operative, **Open Source** ethos.

Netscape: Short for Netscape Communications Inc., the company established by Jim Clark and Marc Andreessen to produce the first commercial Web **browser** and related software. The browser was called **Navigator**, but most people used the word 'Netscape' when referring to the browser, as in 'it looks different in Netscape'. Netscape was bought by AOL in 1999.

Network: A collection of people or (more frequently) computers linked together to share data and information.

Noise: Any signal in a communications network that is considered extraneous.

Open Source: Software circulated as **source code** and protected not by conventional copyright licensing, but by **copyleft**.

Open Source Movement: An influential group within the programming community with a shared set of beliefs about the moral and technical merits of **source code**, co-operative working and **copyleft** licensing.

Operating System: The program which enables a computer to function. Handles tasks like input and output, managing files and memory, interfacing with networks, security and administration.

Packet: A way of handling data for computer communications. In most network applications, messages (string of bits) are not sent in a continuous stream, but broken up into smaller chunks called packets. In a packet, the data being transported are wrapped up in a kind of digital 'envelope' containing the address to which the packet is being sent, the address of the sending station, a number indicating its place in the sequence of packets which make up the whole of the original message, and information which helps to detect and correct errors in transmission.

Packet-switching: A method of sending data in packets from one part of a network to another without having to open a dedicated connection between sender and receiver and keep it open until the whole of the message is delivered. The opposite of **circuit-switching**.

Paradigm: A term used by Thomas Kuhn to describe an established theoretical framework, generally serving as the basis for a scientific or technical discipline. Typically a paradigm defines what constitutes an acceptable experiment, proof, etc. It also determines what counts as an interesting problem. The term is often loosely used to denote a mindset or established way of thinking. Kuhn used the concept to define what he called 'normal science' – i.e. puzzle-solving conducted within the methodological and theoretical framework of a given paradigm. In the Kuhnian model, intellectual progress usually involves periods of great upheaval which he called 'scientific revolutions'. These are periods in which one paradigm is overthrown by a new one, much as in political revolutions old regimes are overthrown by new ones.

PARC: Short for Xerox Palo Alto Research Center – the advanced research laboratory set up by Xerox Corporation in the late 1960s within which much of the computer technology in use in today's personal computers was invented. The first Director of the Computer Science Laboratory at PARC was Robert A. Taylor, the **IPTO** Director who conceived and funded the **ARPANET** project.

PDP: Generic name for a line of minicomputers manufactured by the Digital Equipment Corporation (**DEC**). The company was founded by Ken Olsen, an MIT researcher who had worked on Whirlwind and the TX line of machines.

Portal: The name given to any Website which regularly attracts large numbers of **WWW** users and it is therefore seen as a kind of 'gateway' on to the Net. The major **search engines** are seen as portals.

Protocol: The conventions governing the dialogue between two machines. For example, the World Wide Web's **HTTP** protocol governs the interactions between a **client** (a user's browser) and a Web **server** when the client requests a Web page.

RAM: Random Access Memory – the working memory of a computer. It's volatile in the sense that the contents of RAM are lost the moment the computer is switched off. Contrasts with **ROM**.

RFC: Request for Comment. The name given by the founders of the **ARPANET**'s Network Working Group to their working papers. These papers define the working **protocols** of the Net and were labelled RFC1, RFC2, etc. They are archived in various sites round the world.

ROM: Read-only memory. Software encoded in hardware – either silicon or disk – which is non-volatile but cannot be altered.

Search engine: A special program running on a computer (or network of computers) which searches the Web, indexing or cataloguing the pages it encounters. In practice the term search engine is given to two different kinds of facility – indexing systems like **AltaVista**, and directories like **Yahoo!**, which use human reviewers to compile directories of sites in various categories.

Server: A computer on a network which provides services to other computers running **client** programs.

Servomechanism: A machine which uses feedback automatically to control its output(s) in such a way that discrepancies between the output(s) and some target variable(s) are minimised.

Signal: A sequence of values of something considered significant. In computers and communications the relevant values represent changes in electrical potential.

Signal-to-noise ratio: The mathematical ratio of the signal power to the noise power in a communications channel. Usually measured in decibels.

Silicon: A naturally occurring substance with interesting electrical properties which enable one to create tiny electronic switches known as transistors. Wafers of silicon are the building blocks for almost all current computing devices.

Silicon Valley: An area of northern California about twenty-five miles south of San Francisco centred on the town of Palo Alto and the Stanford University campus. So called because it was the original location of many of the companies and laboratories which created the personal computer industry – Hewlett-Packard, Intel, Apple, Xerox **PARC**, to name just four.

Software: Generic term for computer programs. Contrasts with **hardware**.

Source code: What the programmer writes – i.e. the text of a computer program, composed in a high-level language such as C or BASIC.

Surfing: **Browsing** with one's brain in neutral.

TCP/IP: Stands for 'Transmission Control Protocol/Internet Protocol'. The core of the family of protocols which enable the **Internet** to work by linking thousands of disparate networks into an apparently seamless whole. IP governs the way computers are addressed, while TCP governs the transport of packets.

TELNET: The first of the **ARPANET protocols**. Enables a user to log on to a remote machine over the **Internet** and use its services.

Time-sharing: A method of operating large computers which enables many people to use the computer simultaneously. It works by giving each user a few milliseconds of computing time, but by cycling quickly and continuously between users gives each the illusion of having the machine to him/herself.

UNIX: A multi-user, multi-tasking operating system developed by Ken Thompson and Dennis Ritchie at **Bell Labs** in the 1970s. Subsequently the operating system (OS) of choice of computer science departments world wide, and the OS used by many of the most sophisticated computer systems available. UNIX was always owned by AT&T but in the beginning was distributed free. Later it became a commercial product, a development which sparked Linus Torvalds, a

Finnish student, to create a free clone of UNIX, which he christened **Linux**.

USENET: Often described as 'the world's largest conversation'. A system of over 20,000 specialist discussion groups which use e-mail to post messages. USENET's origins lie in the desire of **UNIX** users to share information about bugs, new releases, etc., but later grew to incorporate almost every conceivable human interest.

UUCP: Short for 'UNIX to UNIX Copy Protocol'. A method of moving files from one **UNIX** machine to another via e-mail. The technology which enabled the creation of USENET.

Virtual: Not real. In computing generally used to denote a process in which something is simulated using computational processes. Thus 'virtual memory' is a technique for using part of a computer's hard disk as if it were **RAM**.

WAN: Wide Area Network – a computer network linking computers in geographically dispersed locations. Contrasts with **LAN**.

World Wide Web: A **hypermedia** system linking text, graphics, sound and video on computers distributed all over the world. Written as three separate words and sometimes abbreviated to W3.

WWW: See **World Wide Web**.

Xanadu: A vast software project aimed at realising Ted Nelson's dream of a 'docuverse' or universe of hyperlinked documents.

YAHOO!: A Web directory set up by two Stanford graduate students, David Filo and Jerry Yang. Now a major **search engine** and **portal** site.

Index